FINANCIAL ENGINEERING
ADVANCED BACKGROUND SERIES

FE PRESS
New York

Financial Engineering Advanced Background Series

Published or forthcoming

1. A Primer for the Mathematics of Financial Engineering, Second Edition, by Dan Stefanica. FE Press, 2011

2. Solutions Manual – A Primer for the Mathematics of Financial Engineering, Second Edition, by Dan Stefanica. FE Press, 2011

3. A Probability Primer for Mathematical Finance, by Elena Kosygina

4. Numerical Linear Algebra Methods for Financial Engineering Applications, by Dan Stefanica

5. Differential Equations with Numerical Methods for Financial Engineering, by Dan Stefanica and Tai-Ho Wang

SOLUTIONS MANUAL

A Primer for the Mathematics of Financial Engineering

Second Edition

DAN STEFANICA

Baruch College
City University of New York

FE Press
New York

FE PRESS
New York

www.fepress.org

©Dan Stefanica 2011

All rights reserved. No part of this publication may be reproduced, stored in a retrieval system, or transmitted, in any form or by any means, electronic, mechanical, photocopying, recording, or otherwise, without the prior written permission of the publisher.

This edition first published 2011

Printed in the United States of America

ISBN-13 978-0-9797576-1-7
ISBN-10 0-9797576-1-4

Cover art and design by Max Rumyantsev

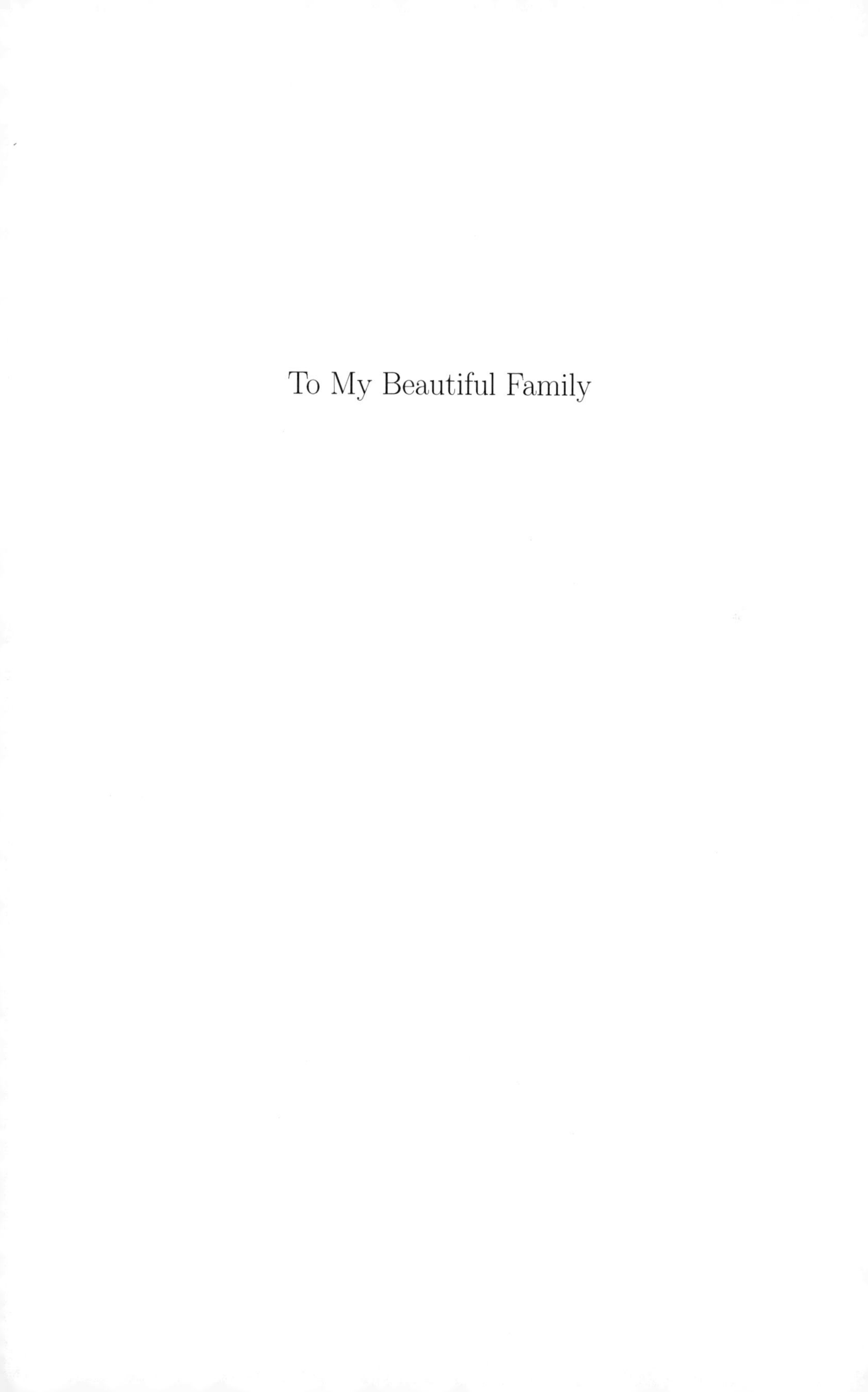

To My Beautiful Family

Contents

Preface to the Second Edition

This Solutions Manual contains solutions to every exercise from the Second Edition of "A Primer for the Mathematics of Financial Engineering".

The Supplemental Exercises from the First Edition are included in the Second Edition of the Math Primer, and their solutions appear in this Solutions Manual. New exercises were added to the Second Edition of the Math Primer, including exercises on the new topics covered therein, for a total of over 170 exercises.

Studying the material from the Math Primer and using the Solutions Manual as a companion is a time effective way to ensure that the reader can solve every exercise, therefore gaining an appropriate depth of understanding of the material from the book.

As Director of the Financial Engineering Program at Baruch College[1] since its inception in 2002, the author has had the privilege to interact with generations of students who were exceptional not only in terms of knowledge and ability, but foremost as very special friends and colleagues. The community that evolved around the alumni, students, and faculty of the program embodies the best things about the Baruch MFE Program: the friendliness and mutual support of everyone involved, in a highly competitive and ultimately very rewarding environment.

"A Primer for the Mathematics of Financial Engineering" and this Solutions Manual are the first books in the Financial Engineering Advanced Background Series, devoted to covering the mathematical background needed for financial engineering applications. While progress was slower than expected, books on Numerical Linear Algebra, Probability, and Differential Equations are forthcoming.

Dan Stefanica

New York, 2011

[1] Baruch MFE Program web page: http://mfe.baruch.cuny.edu

Preface to the First Edition

The addition of this Solutions Manual to "A Primer for the Mathematics of Financial Engineering" offers the reader the opportunity to undertake rigorous self–study of the mathematical topics presented in the Math Primer, with the goal of achieving a deeper understanding of the financial applications therein.

Every exercise from the Math Primer is solved in detail in the Solutions Manual.

Over 50 new exercises are included, and complete solutions to these supplemental exercises are provided. Many of these exercises are quite challenging and offer insight that promises to be most useful in further financial engineering studies as well as job interviews.

Using the Solution Manual as a companion to the Math Primer, the reader will be able to not only bridge any gaps in knowledge but will also glean a more advanced perspective on financial applications by studying the supplemental exercises and their solutions.

The Solutions Manual will be an important resource for prospective financial engineering graduate students. Studying the material from the Math Primer in tandem with the Solutions Manual would provide the solid mathematical background required for successful graduate studies.

The author has been the Director of the Baruch College MFE Program since its inception in 2002. Over 90 percent of the graduates of the Baruch MFE Program are currently employed in the financial industry.

"A Primer for the Mathematics of Financial Engineering" and this Solutions Manual are the first to appear in the Financial Engineering Advanced Background Series. Books on Numerical Linear Algebra, on Probability, and on Differential Equations for financial engineering applications are forthcoming.

Dan Stefanica

New York, 2008

Acknowledgments

I have been the Director of the Financial Engineering Masters Program at Baruch College for many wonderful years. It is a privilege to have the opportunity to work with, and to educate so many talented students, and seeing a strong community develop around the MFE program is incredibly rewarding.

The book "A Primer for the Mathematics of Financial Engineering", and the Solutions Manual, were developed as textbooks for the Advanced Calculus with Financial Applications refresher seminars that I have been teaching since 2004, at first, exclusively to students of the Baruch MFE Program, and, since recently, to a much wider audience. Many of the exercises from this Solutions Manual first arose in these refresher seminars. I am grateful to everyone who took these seminars for the creative energy that goes together with teaching these seminars

Special thanks go to those who supported the proofreading effort of the solutions Manual: for the First Edition, to Barnett Feingold, Chuan Yuan-Huang, Aditya Chitral, Hao He, Weidong Huang, Eugene Krel, Shlomo Ben Shoshan, Mark Su, Shuwen Zhao, and Stefan Zota; for the Second Edition to Nick Chiang, Joe Ejzak, Daniel Forman, and Simon Zheng.

The art for the book cover was again designed by Max Rumyantsev, and Andy Nguyen continued to lend his tremendous support to the fepress.org website and on QuantNetwork. I am indebted to them for all their help.

My wonderful family, to whom this book is dedicated, continues to give sense to my work and make everything worthwhile. Thank you kindly and lovingly!

Dan Stefanica

New York, 2010

Chapter 1

Calculus review. Options. Put–Call parity.

1.1 Exercises

1. Use integration by parts to compute

$$\int \ln(x)\, dx.$$

2. Compute

$$\int \frac{1}{x \ln(x)}\, dx.$$

   Hint: Use the substitution $u = \ln(x)$.

3. Show that $(\tan x)' = 1/(\cos x)^2$ and conclude that

$$\int \frac{1}{1+x^2}\, dx \;=\; \arctan(x) \;+\; C.$$

   Note: The antiderivative of a rational function is often computed using the substitution $x = \tan\left(\frac{z}{2}\right)$.

4. Compute

$$\int x^n \ln(x)\, dx.$$

5. Compute

$$\int x^n e^x\, dx.$$

6. Compute
$$\int (\ln(x))^n \, dx.$$

7. Show that
$$\left(1+\frac{1}{x}\right)^x < e < \left(1+\frac{1}{x}\right)^{x+1}, \quad \forall \, x \geq 1.$$

8. Use l'Hôpital's rule to show that the following two Taylor approximations hold when x is close to 0:
$$\sqrt{1+x} \approx 1 + \frac{x}{2};$$
$$e^x \approx 1 + x + \frac{x^2}{2}.$$
In other words, show that the following limits exist and are constant:
$$\lim_{x\to 0} \frac{\sqrt{1+x} - \left(1+\frac{x}{2}\right)}{x^2} \quad \text{and} \quad \lim_{x\to 0} \frac{e^x - \left(1+x+\frac{x^2}{2}\right)}{x^3}.$$

9. Compute the following limits:
$$\text{(i)} \lim_{x\to\infty} \frac{1}{\sqrt{x^2-4x+1}-x};$$
$$\text{(ii)} \lim_{x\to\infty} \frac{1}{\sqrt{x^2-4x+1}-x+2}.$$

10. Use the definition
$$e = \lim_{x\to\infty} \left(1+\frac{1}{x}\right)^x$$
to show that
$$\frac{1}{e} = \lim_{x\to\infty} \left(1-\frac{1}{x}\right)^x.$$
Hint: Use the fact that
$$\frac{1}{1+\frac{1}{x}} = \frac{x}{x+1} = 1 - \frac{1}{x+1}.$$

11. Let K, T, σ and r be positive constants, and define the function $g : \mathbb{R} \to \mathbb{R}$ as

$$g(x) = \frac{1}{\sqrt{2\pi}} \int_0^{b(x)} e^{-\frac{y^2}{2}} \, dy,$$

where

$$b(x) = \left(\ln\left(\frac{x}{K}\right) + \left(r + \frac{\sigma^2}{2}\right) T \right) / \left(\sigma\sqrt{T}\right).$$

Compute $g'(x)$.

Note: This function is related to the Delta of a plain vanilla Call option.

12. Let $f(x)$ be a continuous function. Show that

$$\lim_{h \to 0} \frac{1}{2h} \int_{a-h}^{a+h} f(x) \, dx = f(a), \quad \forall \ a \in \mathbb{R}.$$

Note: Let $F(x) = \int f(x) \, dx$. The central finite difference approximation of $F'(a)$ is

$$F'(a) = \frac{F(a+h) - F(a-h)}{2h} + O\left(h^2\right), \qquad (1.1)$$

as $h \to 0$ (if $F^{(3)}(x) = f''(x)$ is continuous). Since $F'(a) = f(a)$, formula (1.1) can be written as

$$f(a) = \frac{1}{2h} \int_{a-h}^{a+h} f(x) \, dx + O(h^2).$$

13. Let

$$f(x) = \frac{1}{\sigma\sqrt{2\pi}} \exp\left(-\frac{(x-\mu)^2}{2\sigma^2} \right).$$

Assume that $g : \mathbb{R} \to \mathbb{R}$ is a continuous function which is uniformly bounded, i.e., there exists a constant C such that $|g(x)| \leq C$ for all $x \in \mathbb{R}$. Then, show that

$$\lim_{\sigma \searrow 0} \int_{-\infty}^{\infty} f(x) g(x) \, dx = g(\mu).$$

Note: The function $f(x)$ is the probability density function of a normal random variable with mean μ and standard deviation σ. This exercise shows that the probability density functions of normal variables with standard deviation going to 0 converges, in the sense of distributions, to the delta function corresponding to the mean μ.

14. Let c_i and t_i, $i = 1 : n$, be positive constants.

(i) Let $f : \mathbb{R} \to \mathbb{R}$ given by

$$f(y) \;=\; \sum_{i=1}^{n} c_i e^{-yt_i}.$$

Compute $f'(y)$ and $f''(y)$.

(ii) Let $g : \mathbb{R} \to \mathbb{R}$ given by

$$g(y) \;=\; \sum_{i=1}^{n} c_i \left(1 + \frac{y}{m}\right)^{-mt_i}.$$

Compute $g'(y)$ and $g''(y)$.

Note: The functions $f(y)$ and $g(y)$ represent the price of a bond with cash flows c_i paid at time t_i as a function of the yield y of the bond, if the yield is compounded continuously, and if the yield is compounded discretely m times a year, respectively.

15. Let $f : \mathbb{R}^3 \to \mathbb{R}$ given by

$$f(x) \;=\; 2x_1^2 - x_1x_2 + 3x_2x_3 - x_3^2,$$

where $x = (x_1, x_2, x_3)$.

(i) Compute the gradient and Hessian of the function $f(x)$ at the point $a = (1, -1, 0)$, i.e., compute $Df(1, -1, 0)$ and $D^2f(1, -1, 0)$.

(ii) Show that

$$f(x) \;=\; f(a) + Df(a)\,(x - a) + \frac{1}{2}\,(x - a)^t\,D^2f(a)\,(x - a). \qquad (1.2)$$

Here, x, a, and $x - a$ are 3×1 column vectors, i.e.,

$$x \;=\; \begin{pmatrix} x_1 \\ x_2 \\ x_3 \end{pmatrix}; \quad a \;=\; \begin{pmatrix} 1 \\ -1 \\ 0 \end{pmatrix}; \quad x - a \;=\; \begin{pmatrix} x_1 - 1 \\ x_2 + 1 \\ x_3 \end{pmatrix}.$$

Note: Formula (1.2) is the quadratic Taylor approximation of $f(x)$ around the point a. Since $f(x)$ is a second order polynomial, the quadratic Taylor approximation of $f(x)$ is exact.

16. Let

$$u(x,t) = \frac{1}{\sqrt{4\pi t}} e^{-\frac{x^2}{4t}}, \quad \text{for } t > 0, \ x \in \mathbb{R}.$$

Compute $\frac{\partial u}{\partial t}$ and $\frac{\partial^2 u}{\partial x^2}$, and show that

$$\frac{\partial u}{\partial t} = \frac{\partial^2 u}{\partial x^2}.$$

Note: This exercise shows that the function $u(x,t)$ is a solution of the heat equation. In fact, $u(x,t)$ is the fundamental solution of the heat equation, and is used in the PDE derivation of the Black–Scholes formula for pricing European plain vanilla options.

Also, note that $u(x,t)$ is the same as the density function of a normal variable with mean 0 and variance $2t$.

17. Show that the values of a plain vanilla put option and of a plain vanilla call option with the same maturity and strike, and on the same underlying asset, are equal if and only if the strike is equal to the forward price.

18. Consider a portfolio with the following positions:
   - long one call option with strike $K_1 = 30$;
   - short two call options with strike $K_2 = 35$;
   - long one call option with strike $K_3 = 40$.

All options are on the same underlying asset and have maturity T. Draw the payoff diagram at maturity of the portfolio, i.e., plot the value of the portfolio $V(T)$ at maturity as a function of $S(T)$, the price of the underlying asset at time T.

Note: This is a butterfly spread. A trader takes a long position in a butterfly spread if the price of the underlying asset at maturity is expected to be in the $K_1 \leq S(T) \leq K_3$ range.

19. Draw the payoff diagram at maturity of a bull spread with a long position in a call with strike 30 and short a call with strike 35, and of a bear spread with long a put of strike 20 and short a put of strike 15.

20. The prices of three call options with strikes 45, 50, and 55, on the same underlying asset and with the same maturity, are \$4, \$6, and \$9, respectively. Create a butterfly spread by going long a 45–call and a

55–call, and shorting two 50–calls. What are the payoff and the P&L at maturity of the butterfly spread? When would the butterfly spread be profitable? Assume, for simplicity, that interest rates are zero.

21. Which of the following two portfolios would you rather hold:

    - Portfolio 1: Long one call option with strike $K = X - 5$ and long one call option with strike $K = X + 5$;
    - Portfolio 2: Long two call options with strike $K = X$?

    (All options are on the same asset and have the same maturity.)

22. A stock with spot price \$42 pays dividends continuously at a rate of 3%. The four months put and call options with strike 40 on this asset are trading at \$2 and \$4, respectively. The risk-free rate is constant and equal to 5% for all times. Show that the Put-Call parity is not satisfied and explain how would you take advantage of this arbitrage opportunity.

23. The bid and ask prices for a six months European call option with strike 40 on a non–dividend–paying stock with spot price 42 are \$5 and \$5.5, respectively. The bid and ask prices for a six months European put option with strike 40 on the same underlying asset are \$2.75 and \$3.25, respectively. Assume that the risk free rate is equal to 0. Is there an arbitrage opportunity present?

24. Denote by C_{bid} and C_{ask}, and by P_{bid} and P_{ask}, respectively, the bid and ask prices for a plain vanilla European call and for a plain vanilla European put option, both with the same strike K and maturity T, and on the same underlying asset with spot price S and paying dividends continuously at rate q. Assume that the risk–free interest rates are constant equal to r. Find necessary and sufficient no–arbitrage conditions for C_{bid}, C_{ask}, P_{bid}, and P_{ask}.

25. You expect that an asset with spot price \$35 will trade in the \$40–\$45 range in one year. One year at–the–money calls on the asset can be bought for \$4. To act on the expected stock price appreciation, you decide to either buy the asset, or to buy ATM calls. Which strategy is better, depending on where the asset price will be in a year?

26. Create a portfolio with the following payoff at time T:

$$V(T) = \begin{cases} 2S(T), & \text{if } 0 \leq S(T) < 20; \\ 60 - S(T), & \text{if } 20 \leq S(T) < 40; \\ S(T) - 20, & \text{if } 40 \leq S(T), \end{cases}$$

where $S(T)$ is the spot price at time T of a given asset. Use plain vanilla options with maturity T as well as cash positions and positions in the asset itself. Assume, for simplicity, that the asset does not pay dividends and that interest rates are zero.

27. A derivative security pays a cash amount c if the spot price of the underlying asset at maturity is between K_1 and K_2, where $0 < K_1 < K_2$, and expires worthless otherwise. How do you synthesize this derivative security (i.e., how do you recreate its payoff almost exactly) using plain vanilla call options?

28. Call options with strikes 100, 120, and 130 on the same underlying asset and with the same maturity are trading for 8, 5, and 3, respectively (there is no bid–ask spread). Is there an arbitrage opportunity present? If yes, how can you make a riskless profit?

29. Call options on the same underlying asset and with the same maturity, with strikes $K_1 < K_2 < K_3$, are trading for C_1, C_2 and C_3, respectively (no Bid–Ask spread), with $C_1 > C_2 > C_3$. Find necessary and sufficient conditions on the prices C_1, C_2 and C_3 such that no–arbitrage exists corresponding to a portfolio made of positions in the three options.

30. The risk free rate is 8% compounded continuously and the dividend yield of a stock index is 3%. The index is at 12,000 and the futures price of a contract deliverable in three months is 12,100. Is there an arbitrage opportunity, and how do you take advantage of it?

1.2 Solutions to Chapter 1 Exercises

Problem 1: Compute

$$\int \ln(x)\, dx.$$

Solution: Using integration by parts, we find that

$$\begin{aligned}\int \ln(x)\, dx &= \int (x)' \ln(x)\, dx = x\ln(x) - \int x(\ln(x))'\, dx \\ &= x\ln(x) - \int 1\, dx = x\ln(x) - x + C. \quad \square\end{aligned}$$

Problem 2: Compute

$$\int \frac{1}{x\ln(x)}\, dx$$

by using the substitution $u = \ln(x)$.

Solution: Let $u = \ln(x)$. Then $du = \frac{dx}{x}$ and therefore

$$\int \frac{1}{x\ln(x)}\, dx = \int \frac{1}{u}\, du = \ln(|u|) = \ln(|\ln(x)|) + C. \quad \square$$

Problem 3: Show that $(\tan x)' = 1/(\cos x)^2$ and conclude that

$$\int \frac{1}{1+x^2}\, dx = \arctan(x) + C.$$

Solution: Using the Quotient Rule, we find that

$$\begin{aligned}(\tan x)' &= \left(\frac{\sin x}{\cos x}\right)' = \frac{(\sin x)'\cos x - \sin x(\cos x)'}{(\cos x)^2} \\ &= \frac{(\cos x)^2 + (\sin x)^2}{(\cos x)^2} = \frac{1}{(\cos x)^2}.\end{aligned} \tag{1.3}$$

To prove that $\int \frac{1}{1+x^2} dx = \arctan(x)$, we will show that

$$(\arctan(x))' = \frac{1}{1+x^2}.$$

Let $f(x) = \tan x$. Then $\arctan(x) = f^{-1}(x)$. Recall that

$$\left(f^{-1}(x)\right)' = \frac{1}{f'(f^{-1}(x))}$$

and note that $f'(x) = (\tan x)' = \frac{1}{(\cos x)^2}$; cf. (1.3). Therefore,

$$(\arctan(x))' = (\cos(f^{-1}(x)))^2 = (\cos(\arctan(x)))^2. \tag{1.4}$$

Let $\alpha = \arctan(x)$. Then $\tan(\alpha) = x$. It is easy to see that

$$x^2 + 1 = \frac{1}{(\cos(\alpha))^2},$$

since $(\sin(\alpha))^2 + (\cos(\alpha))^2 = 1$. Thus,

$$(\cos(\arctan(x)))^2 \;=\; (\cos(\alpha))^2 \;=\; \frac{1}{x^2+1}. \tag{1.5}$$

From (1.4) and (1.5), we conclude that

$$(\arctan(x))' \;=\; \frac{1}{x^2+1},$$

and therefore that

$$\int \frac{1}{1+x^2}\, dx \;=\; \arctan(x) \;+\; C.$$

We note that the antiderivative of a rational function is often computed using the substitution $x = \tan\left(\frac{z}{2}\right)$.

For example, to compute $\int \frac{1}{1+x^2} dx$ using the substitution $x = \tan\left(\frac{z}{2}\right)$, note that

$$dx \;=\; \frac{d}{dz}\left(\tan\left(\frac{z}{2}\right)\right) dz \;=\; \frac{1}{2(\cos\left(\frac{z}{2}\right))^2} dz.$$

Then

$$\begin{aligned}
\int \frac{1}{1+x^2}\, dx \;&=\; \int \frac{1}{1+(\tan\left(\frac{z}{2}\right))^2} \cdot \frac{1}{2(\cos\left(\frac{z}{2}\right))^2}\, dz \\
&=\; \int \frac{\left(\cos\left(\frac{z}{2}\right)\right)^2}{(\sin(\alpha))^2 + (\cos(\alpha))^2} \cdot \frac{1}{2(\cos\left(\frac{z}{2}\right))^2}\, dz \\
&=\; \int \frac{1}{2}\, dz \;=\; \frac{z}{2} = \arctan(x) \;+\; C. \quad \square
\end{aligned}$$

Problem 4: Compute

$$\int x^n \ln(x)\, dx.$$

Solution: If $n \neq -1$, we use integration by parts and find that

$$\begin{aligned} \int x^n \ln(x)\, dx &= \frac{x^{n+1}}{n+1}\ln(x) - \int \frac{x^{n+1}}{n+1} \cdot \frac{1}{x}\, dx \\ &= \frac{x^{n+1}\ln(x)}{n+1} - \frac{1}{n+1}\int x^n\, dx \\ &= \frac{x^{n+1}\ln(x)}{n+1} - \frac{x^{n+1}}{(n+1)^2} + C. \end{aligned}$$

For $n = -1$, we obtain that

$$\int \frac{\ln(x)}{x}\, dx = \frac{(\ln(x))^2}{2} + C. \quad \square$$

Problem 5: Compute

$$\int x^n e^x\, dx.$$

Solution: For every integer $n \geq 0$, let

$$f_n(x) = \int x^n e^x\, dx.$$

By using integration by parts, we find that

$$f_n(x) = x^n e^x - n\int x^{n-1} e^x\, dx,$$

which can be written as

$$f_n(x) = x^n e^x - n f_{n-1}(x), \quad \forall\, n \geq 1. \tag{1.6}$$

Note that[1]

$$f_0(x) = \int e^x dx = e^x.$$

By letting $n = 1$ in (1.6), we obtain that

$$f_1(x) = xe^x - f_0(x) = (x-1)e^x.$$

By letting $n = 2$ in (1.6), we obtain that

$$f_2(x) = x^2 e^x - 2f_1(x) = (x^2 - 2x + 2)e^x.$$

[1]To avoid confusions, we will not add a constant C when writing down the formulas for $f_0(x)$, $f_1(x)$, $f_2(x)$, and $f_3(x)$.

By letting $n = 3$ in (1.6), we obtain that

$$f_3(x) \;=\; x^3e^x - 3f_2(x) \;=\; (x^3 - 3x^2 + 6x - 6)e^x.$$

The following general formula can be proved by induction:

$$\int x^n e^x \, dx \;=\; f_n(x) \;=\; \left(\sum_{k=0}^{n} x^k \frac{(-1)^{n-k} n!}{k!}\right) e^x \;+\; C, \quad \forall\, n \geq 0. \qquad \square$$

Problem 6: Compute

$$\int (\ln(x))^n \, dx.$$

Solution: For every integer $n \geq 0$, let

$$f_n(x) \;=\; \int (\ln(x))^n \, dx.$$

By using integration by parts, it is easy to see that, for any $n \geq 1$,

$$\begin{aligned}
\int (\ln(x))^n \, dx \;&=\; x(\ln(x))^n \;-\; \int x\left((\ln(x))^n\right)' \, dx \\
&=\; x(\ln(x))^n \;-\; \int x \cdot n(\ln(x))^{n-1} \cdot (\ln(x))' \, dx \\
&=\; x(\ln(x))^n \;-\; \int x \cdot n(\ln(x))^{n-1} \cdot \frac{1}{x} \, dx \\
&=\; x(\ln(x))^n - n\int (\ln(x))^{n-1} \, dx,
\end{aligned}$$

and therefore

$$f_n(x) \;=\; x(\ln(x))^n - nf_{n-1}(x), \quad \forall\, n \geq 1. \tag{1.7}$$

Note that

$$f_0(x) \;=\; \int 1dx \;=\; x.$$

By letting $n = 1$ in (1.7), we obtain that

$$f_1(x) \;=\; x\ln(x) - f_0(x) \;=\; x(\ln(x) - 1).$$

By letting $n = 2$ in (1.7), we obtain that

$$f_2(x) \;=\; x(\ln(x))^2 - 2f_1(x) \;=\; x\left((\ln(x))^2 - 2\ln(x) + 2\right).$$

By letting $n = 3$ in (1.7), we obtain that

$$f_3(x) = x(\ln(x))^3 - 3f_2(x) = x\left(\ln(x))^3 - 3(\ln(x))^2 + 6\ln(x) - 6\right).$$

The following general formula can be obtained by induction:

$$\int (\ln(x))^n \, dx = x \sum_{k=0}^{n} \frac{(-1)^{n-k} n!}{k!} (\ln(x))^k + C, \ \forall\, n \geq 0. \ \square$$

Problem 7: Show that

$$\left(1 + \frac{1}{x}\right)^x < e < \left(1 + \frac{1}{x}\right)^{x+1}, \ \forall\, x \geq 1. \tag{1.8}$$

Solution: By taking natural logs on both sides of (1.8), we find that (1.8) is equivalent to

$$x \ln\left(1 + \frac{1}{x}\right) < 1 < (x+1)\ln\left(1 + \frac{1}{x}\right),$$

which can be written as

$$\frac{1}{x+1} < \ln\left(1 + \frac{1}{x}\right) < \frac{1}{x}, \ \forall\, x \geq 1. \tag{1.9}$$

Let

$$f(x) = \frac{1}{x} - \ln\left(1 + \frac{1}{x}\right); \quad g(x) = \ln\left(1 + \frac{1}{x}\right) - \frac{1}{x+1}.$$

Then,

$$f'(x) = -\frac{1}{x^2} + \frac{1}{x(x+1)} = -\frac{1}{x^2(x+1)} < 0;$$

$$g'(x) = -\frac{1}{x(x+1)} + \frac{1}{(x+1)^2} = -\frac{1}{x(x+1)^2} < 0$$

We conclude that both $f(x)$ and $g(x)$ are decreasing functions. Since

$$\lim_{x\to\infty} f(x) = \lim_{x\to\infty} g(x) = 0,$$

it follows that $f(x) > 0$ and $g(x) > 0$ for all $x > 0$, and therefore

$$\frac{1}{x} > \ln\left(1 + \frac{1}{x}\right) > \frac{1}{x+1}, \ \forall\, x > 0,$$

which is what we wanted to show; cf. (1.9). □

Problem 8: Use l'Hôpital's rule to show that the following two Taylor approximations hold when x is close to 0:

$$\begin{aligned}\sqrt{1+x} &\approx 1 + \frac{x}{2};\\ e^x &\approx 1 + x + \frac{x^2}{2}.\end{aligned}$$

In other words, show that the following limits exist and are constant:

$$\lim_{x\to 0} \frac{\sqrt{1+x} - \left(1+\frac{x}{2}\right)}{x^2} \quad \text{and} \quad \lim_{x\to 0} \frac{e^x - \left(1+x+\frac{x^2}{2}\right)}{x^3}.$$

Solution: The numerator and denominator of each limit are differentiated until a finite limit is computed. L'Hôpital's rule can then be applied sequentially to obtain the value of the initial limit:

$$\begin{aligned}\lim_{x\to 0} \frac{\sqrt{1+x} - \left(1+\frac{x}{2}\right)}{x^2} &= \lim_{x\to 0} \frac{\frac{1}{2\sqrt{1+x}} - \frac{1}{2}}{2x}\\ &= \lim_{x\to 0} \frac{-\frac{1}{4(1+x)^{3/2}}}{2}\\ &= -\frac{1}{8}.\end{aligned}$$

We conclude that

$$\sqrt{1+x} = 1 + \frac{x}{2} + O(x^2), \quad \text{as} \quad x \to 0.$$

Similarly,

$$\begin{aligned}\lim_{x\to 0} \frac{e^x - \left(1+x+\frac{x^2}{2}\right)}{x^3} &= \lim_{x\to 0} \frac{e^x - (1+x)}{3x^2}\\ &= \lim_{x\to 0} \frac{e^x - 1}{6x}\\ &= \lim_{x\to 0} \frac{e^x}{6}\\ &= \frac{1}{6},\end{aligned}$$

and therefore

$$e^x = 1 + x + \frac{x^2}{2} + O(x^3) \quad \text{as} \quad x \to 0. \quad \square$$

Problem 9: Compute the following limits:

$$\text{(i)} \lim_{x\to\infty} \frac{1}{\sqrt{x^2-4x+1}-x};$$

$$\text{(ii)} \lim_{x\to\infty} \frac{1}{\sqrt{x^2-4x+1}-x+2}.$$

Solution: (i) By multiplying the denominator of the fraction with its conjugate, it is easy to see that

$$\begin{aligned}
\lim_{x\to\infty} \frac{1}{\sqrt{x^2-4x+1}-x} &= \lim_{x\to\infty} \frac{\sqrt{x^2-4x+1}+x}{(\sqrt{x^2-4x+1}+x)(\sqrt{x^2-4x+1}-x)} \\
&= \lim_{x\to\infty} \frac{\sqrt{x^2-4x+1}+x}{(\sqrt{x^2-4x+1})^2 - x^2} \\
&= \lim_{x\to\infty} \frac{\sqrt{x^2-4x+1}+x}{x^2-4x+1 - x^2} \\
&= \lim_{x\to\infty} \frac{\sqrt{x^2-4x+1}+x}{-4x+1} \\
&= \lim_{x\to\infty} \frac{\sqrt{1-\frac{4}{x}+\frac{1}{x^2}}+1}{-4+\frac{1}{x}} \\
&= \frac{1+1}{-4} \\
&= -\frac{1}{2}.
\end{aligned}$$

(ii) Using a similar method as before, we find that

$$\begin{aligned}
&\lim_{x\to\infty} \frac{1}{\sqrt{x^2-4x+1}-x+2} \\
&= \lim_{x\to\infty} \frac{1}{\sqrt{x^2-4x+1}-(x-2)} \\
&= \lim_{x\to\infty} \frac{\sqrt{x^2-4x+1}+(x-2)}{(\sqrt{x^2-4x+1}+(x-2))(\sqrt{x^2-4x+1}-(x-2))} \\
&= \lim_{x\to\infty} \frac{\sqrt{x^2-4x+1}+(x-2)}{(\sqrt{x^2-4x+1})^2 - (x-2)^2} \\
&= \lim_{x\to\infty} \frac{\sqrt{x^2-4x+1}+(x-2)}{x^2-4x+1 - (x^2-4x+4)} \\
&= \lim_{x\to\infty} \frac{\sqrt{x^2-4x+1}+(x-2)}{-3}
\end{aligned}$$

$$= -\frac{1}{3}\lim_{x\to\infty}\left(\sqrt{x^2-4x+1}+x-2\right)$$
$$= -\infty. \quad \square$$

Problem 10: Use the definition

$$e = \lim_{x\to\infty}\left(1+\frac{1}{x}\right)^x$$

to show that

$$\frac{1}{e} = \lim_{x\to\infty}\left(1-\frac{1}{x}\right)^x.$$

Solution: Note that

$$1-\frac{1}{x} = \frac{x-1}{x} = \frac{1}{\frac{x}{x-1}} = \frac{1}{1+\frac{1}{x-1}}.$$

Then,

$$\lim_{x\to\infty}\left(1-\frac{1}{x}\right)^x = \lim_{x\to\infty}\frac{1}{\left(1+\frac{1}{x-1}\right)^x}$$
$$= \lim_{x\to\infty}\frac{1}{\left(1+\frac{1}{x-1}\right)^{x-1}}\cdot\frac{1}{1+\frac{1}{x-1}} = \frac{1}{e},$$

since

$$\lim_{x\to\infty} 1+\frac{1}{x-1} = 1$$

and

$$\lim_{x\to\infty}\left(1+\frac{1}{x-1}\right)^{x-1} = \lim_{x\to\infty}\left(1+\frac{1}{x}\right)^x = e. \quad \square$$

Problem 11: Let K, T, σ and r be positive constants, and define the function $g:\mathbb{R}\to\mathbb{R}$ as

$$g(x) = \frac{1}{\sqrt{2\pi}}\int_0^{b(x)} e^{-\frac{y^2}{2}}\,dy,$$

where

$$b(x) = \left(\ln\left(\frac{x}{K}\right)+\left(r+\frac{\sigma^2}{2}\right)T\right)/\left(\sigma\sqrt{T}\right). \tag{1.10}$$

Compute $g'(x)$.

Solution: Recall that

$$\frac{d}{dx}\left(\int_{a(x)}^{b(x)} f(y)\,dy\right) = f(b(x))b'(x) - f(a(x))a'(x), \tag{1.11}$$

and note that

$$b'(x) = \frac{1}{x\sigma\sqrt{T}}.$$

Using (1.11) for $a(x) = 0$, $b(x)$ given by (1.10), and $f(y) = \frac{1}{\sqrt{2\pi}}e^{-\frac{y^2}{2}}$, we obtain that

$$\begin{aligned} g'(x) &= \frac{1}{\sqrt{2\pi}}\, e^{-\frac{(b(x))^2}{2}} \cdot b'(x) = \frac{1}{\sqrt{2\pi}} \exp\left(-\frac{(b(x))^2}{2}\right) \frac{1}{x\sigma\sqrt{T}} \\ &= \frac{1}{x\sigma\sqrt{2\pi T}} \exp\left(-\frac{\left(\ln\left(\frac{x}{K}\right) + \left(r + \frac{\sigma^2}{2}\right)T\right)^2}{2\sigma^2 T}\right). \quad \square \end{aligned}$$

Problem 12: Let $f(x)$ be a continuous function. Show that

$$\lim_{h\to 0} \frac{1}{2h} \int_{a-h}^{a+h} f(x)\,dx = f(a), \quad \forall\, a \in \mathbb{R}.$$

Solution: Let $F(x) = \int f(x)\,dx$ be the antiderivative of $f(x)$. From the Fundamental Theorem of Calculus, it follows that

$$\frac{1}{2h}\int_{a-h}^{a+h} f(x)\,dx = \frac{F(a+h) - F(a-h)}{2h}.$$

Using l'Hôpital's rule and the fact that $F'(x) = f(x)$, we find that

$$\begin{aligned} \lim_{h\to 0} \frac{1}{2h}\int_{a-h}^{a+h} f(x)\,dx &= \lim_{h\to 0} \frac{F(a+h) - F(a-h)}{2h} \\ &= \lim_{h\to 0} \frac{f(a+h) + f(a-h)}{2} \\ &= f(a), \end{aligned}$$

since $f(x)$ is a continuous function. $\square$

Problem 13: Let

$$f(x) = \frac{1}{\sigma\sqrt{2\pi}} \exp\left(-\frac{(x-\mu)^2}{2\sigma^2}\right).$$

Assume that $g : \mathbb{R} \to \mathbb{R}$ is a uniformly bounded[2] continuous function, i.e., assume that there exists a constant C such that $|g(x)| \leq C$ for all $x \in \mathbb{R}$. Show that

$$\lim_{\sigma \searrow 0} \int_{-\infty}^{\infty} f(x)g(x)\, dx \;=\; g(\mu).$$

Solution: Using the change of variables $y = \frac{x-\mu}{\sigma}$, we find that

$$\begin{aligned} \int_{-\infty}^{\infty} f(x)g(x)\, dx &= \frac{1}{\sigma\sqrt{2\pi}} \int_{-\infty}^{\infty} g(x) \exp\left(-\frac{(x-\mu)^2}{2\sigma^2}\right) dx \\ &= \frac{1}{\sqrt{2\pi}} \int_{-\infty}^{\infty} g(\mu + \sigma y)\, e^{-\frac{y^2}{2}}\, dy. \end{aligned} \tag{1.12}$$

Recall that

$$\frac{1}{\sqrt{2\pi}} \int_{-\infty}^{\infty} e^{-\frac{y^2}{2}}\, dy \;=\; 1, \tag{1.13}$$

since, e.g., the function $\frac{1}{\sqrt{2\pi}} e^{-\frac{y^2}{2}}$ is the probability density function of the standard normal variable. From (1.12) and (1.13) we obtain that

$$g(\mu) - \int_{-\infty}^{\infty} f(x)g(x)\, dx \;=\; \frac{1}{\sqrt{2\pi}} \int_{-\infty}^{\infty} (g(\mu) - g(\mu + \sigma y))\, e^{-\frac{y^2}{2}}\, dy. \tag{1.14}$$

Our goal is to show that the right hand side of (1.14) goes to 0 as $\sigma \searrow 0$.

Since $g(x)$ is a continuous function, it follows that, for any $\epsilon > 0$, there exists $\delta_1(\epsilon) > 0$ such that

$$|g(\mu) - g(x)| < \epsilon, \;\; \forall\, |x - \mu| < \delta_1(\epsilon). \tag{1.15}$$

Using the fact that the integral (1.13) exists and is finite, we obtain that, for any $\epsilon > 0$, there exists $\delta_2(\epsilon) > 0$ such that

$$\frac{1}{\sqrt{2\pi}} \int_{-\infty}^{-\delta_2(\epsilon)} e^{-\frac{y^2}{2}}\, dy \;+\; \frac{1}{\sqrt{2\pi}} \int_{\delta_2(\epsilon)}^{\infty} e^{-\frac{y^2}{2}}\, dy \;<\; \epsilon. \tag{1.16}$$

Since $|g(x)| \leq C$ for all $x \in \mathbb{R}$, it follows from (1.16) that

$$\begin{aligned} &\frac{1}{\sqrt{2\pi}} \int_{-\infty}^{-\delta_2(\epsilon)} |g(\mu) - g(\mu + \sigma y)|\, e^{-\frac{y^2}{2}}\, dy \\ &+ \frac{1}{\sqrt{2\pi}} \int_{\delta_2(\epsilon)}^{\infty} |g(\mu) - g(\mu + \sigma y)|\, e^{-\frac{y^2}{2}}\, dy \;<\; 2C\epsilon. \end{aligned} \tag{1.17}$$

[2]The uniform boundedness condition was chosen for simplicity, and it can be relaxed, e.g., to functions which have polynomial growth at infinity.

It is easy to see that, if $\sigma < \frac{\delta_1(\epsilon)}{\delta_2(\epsilon)}$, then

$$|(\mu + \sigma y) - \mu| = \sigma|y| < \delta_1(\epsilon)\frac{|y|}{\delta_2(\epsilon)} \leq \delta_1(\epsilon), \quad \forall\, y \in [-\delta_2(\epsilon), \delta_2(\epsilon)]. \tag{1.18}$$

Then, from (1.15) and (1.18) we find that

$$|g(\mu) - g(\mu + \sigma y)| \;<\; \epsilon, \quad \forall\, y \in [-\delta_2(\epsilon), \delta_2(\epsilon)], \tag{1.19}$$

and therefore

$$\frac{1}{\sqrt{2\pi}} \int_{-\delta_2(\epsilon)}^{\delta_2(\epsilon)} |g(\mu) - g(\mu + \sigma y)| \; e^{-\frac{y^2}{2}} \, dy \;<\; \epsilon. \tag{1.20}$$

From (1.14), (1.17), and (1.20), it follows that, for any $\epsilon > 0$, there exist $\delta_1(\epsilon) > 0$ and $\delta_2(\epsilon) > 0$ such that, if $\sigma < \frac{\delta_1(\epsilon)}{\delta_2(\epsilon)}$, then

$$\begin{aligned} \left| g(\mu) - \int_{-\infty}^{\infty} f(x)g(x)\, dx \right| \;&\leq\; \frac{1}{\sqrt{2\pi}} \int_{-\infty}^{\infty} |g(\mu) - g(\mu + \sigma y)| \; e^{-\frac{y^2}{2}} \, dy \\ &<\; (2C + 1)\epsilon. \end{aligned}$$

We conclude, by definition, that

$$\lim_{\sigma \searrow 0} \int_{-\infty}^{\infty} f(x)g(x)\, dx \;=\; g(\mu). \quad \square$$

Problem 14: Let c_i and t_i, $i = 1 : n$, be positive constants.
(i) Let $f : \mathbb{R} \to \mathbb{R}$ given by

$$f(y) \;=\; \sum_{i=1}^{n} c_i e^{-yt_i}.$$

Compute $f'(y)$ and $f''(y)$.
(ii) Let $g : \mathbb{R} \to \mathbb{R}$ given by

$$g(y) \;=\; \sum_{i=1}^{n} c_i \left(1 + \frac{y}{m}\right)^{-mt_i}.$$

Compute $g'(y)$ and $g''(y)$.

Solution: (i) Note that

$$\begin{aligned} \left(e^{-yt_i}\right)' \;&=\; \frac{d}{dy}\left(e^{-yt_i}\right) \;=\; -\,t_i e^{-yt_i}; \\ \left(e^{-yt_i}\right)'' \;&=\; \frac{d}{dy}\left(-t_i e^{-yt_i}\right) \;=\; t_i^2 e^{-yt_i}. \end{aligned}$$

Then,

$$
\begin{aligned}
f'(y) &= -\sum_{i=1}^{n} c_i t_i e^{-yt_i}; \\
f''(y) &= \sum_{i=1}^{n} c_i t_i^2 e^{-yt_i}.
\end{aligned}
$$

(ii) Using Chain Rule, we obtain that

$$
\begin{aligned}
\left(\left(1+\frac{y}{m}\right)^{-mt_i}\right)' &= \left(1+\frac{y}{m}\right)^{-mt_i-1} \cdot (-mt_i) \cdot \frac{1}{m} \\
&= -t_i \left(1+\frac{y}{m}\right)^{-mt_i-1}; \\
\left(\left(1+\frac{y}{m}\right)^{-mt_i}\right)'' &= -t_i \left(1+\frac{y}{m}\right)^{-mt_i-2} \cdot (-mt_i-1) \cdot \frac{1}{m} \\
&= t_i \left(t_i + \frac{1}{m}\right) \left(1+\frac{y}{m}\right)^{-mt_i-2}.
\end{aligned}
$$

Then,

$$
\begin{aligned}
g'(y) &= -\sum_{i=1}^{n} c_i t_i \left(1+\frac{y}{m}\right)^{-mt_i-1}; \\
g''(y) &= \sum_{i=1}^{n} c_i t_i \left(t_i + \frac{1}{m}\right) \left(1+\frac{y}{m}\right)^{-mt_i-2}. \quad \square
\end{aligned}
$$

Problem 15: Let $f : \mathbb{R}^3 \to \mathbb{R}$ given by

$$f(x) = 2x_1^2 - x_1 x_2 + 3x_2 x_3 - x_3^2,$$

where $x = (x_1, x_2, x_3)$.

(i) Compute the gradient and Hessian of the function $f(x)$ at the point $a = (1, -1, 0)$, i.e., compute $Df(1, -1, 0)$ and $D^2 f(1, -1, 0)$.

(ii) Show that

$$f(x) = f(a) + Df(a)\,(x-a) + \frac{1}{2}\,(x-a)^t\, D^2 f(a)\,(x-a).$$

Here, x, a, and $x - a$ are 3×1 column vectors, i.e.,

$$x = \begin{pmatrix} x_1 \\ x_2 \\ x_3 \end{pmatrix}; \quad a = \begin{pmatrix} 1 \\ -1 \\ 0 \end{pmatrix}; \quad x - a = \begin{pmatrix} x_1 - 1 \\ x_2 + 1 \\ x_3 \end{pmatrix}.$$

Solution: (i) Recall that

$$\begin{aligned} Df(x) &= \left(\frac{\partial f}{\partial x_1}(x) \;\; \frac{\partial f}{\partial x_2}(x) \;\; \frac{\partial f}{\partial x_n}(x) \right) \\ &= (4x_1 - x_2, \;\; -x_1 + 3x_3, \;\; 3x_2 - 2x_3)\,; \\ D^2 f(x) &= \begin{pmatrix} \frac{\partial^2 f}{\partial x_1^2}(x) & \frac{\partial^2 f}{\partial x_2 \partial x_1}(x) & \frac{\partial^2 f}{\partial x_3 \partial x_1}(x) \\ \frac{\partial^2 f}{\partial x_1 \partial x_2}(x) & \frac{\partial^2 f}{\partial x_2^2}(x) & \frac{\partial^2 f}{\partial x_3 \partial x_2}(x) \\ \frac{\partial^2 f}{\partial x_1 \partial x_3}(x) & \frac{\partial^2 f}{\partial x_2 \partial x_3}(x) & \frac{\partial^2 f}{\partial x_3^2}(x) \end{pmatrix} = \begin{pmatrix} 4 & -1 & 0 \\ -1 & 0 & 3 \\ 0 & 3 & -2 \end{pmatrix}. \end{aligned}$$

Then,

$$f(a) = f(1,-1,0) = 3 \tag{1.21}$$

$$Df(a) = Df(1,-1,0) = (5, \;\; -1, \;\; -3); \tag{1.22}$$

$$D^2 f(a) = D^2 f(1,-1,0) = \begin{pmatrix} 4 & -1 & 0 \\ -1 & 0 & 3 \\ 0 & 3 & -2 \end{pmatrix}. \tag{1.23}$$

(ii) We substitute the values from (1.21), (1.22) and (1.23) for $f(a)$, $Df(a)$ and $D^2 f(a)$, respectively, in the expression $f(a) + Df(a)\,(x-a) + \frac{1}{2}\,(x-a)^t\, D^2 f(a)\,(x-a)$ and obtain that

$$\begin{aligned} & f(a) + Df(a)\,(x-a) + \frac{1}{2}\,(x-a)^t\, D^2 f(a)\,(x-a) \\ = \;& 3 \;+\; (5,-1,-3) \begin{pmatrix} x_1 - 1 \\ x_2 + 1 \\ x_3 \end{pmatrix} \\ & + \frac{1}{2}\,(x_1 - 1, x_2 + 1, x_3) \begin{pmatrix} 4 & -1 & 0 \\ -1 & 0 & 3 \\ 0 & 3 & -2 \end{pmatrix} \begin{pmatrix} x_1 - 1 \\ x_2 + 1 \\ x_3 \end{pmatrix} \\ = \;& 3 \;+\; (5x_1 - x_2 - 3x_3 - 6) \\ & + \left(2x_1^2 - 5x_1 - x_1 x_2 + x_2 + 3x_2 x_3 + 3x_3 - x_3^2 + 3\right) \\ = \;& 2x_1^2 - x_1 x_2 + 3x_2 x_3 - x_3^2 \\ = \;& f(x). \;\; \square \end{aligned}$$

Problem 16: Let

$$u(x,t) = \frac{1}{\sqrt{4\pi t}}\, e^{-\frac{x^2}{4t}}, \quad \text{for } \; t > 0, \; x \in \mathbb{R}.$$

Compute $\frac{\partial u}{\partial t}$ and $\frac{\partial^2 u}{\partial x^2}$, and show that

$$\frac{\partial u}{\partial t} = \frac{\partial^2 u}{\partial x^2}.$$

Solution: By direct computation and using the Product Rule, we find that

$$\begin{aligned}
\frac{\partial u}{\partial t} &= -\frac{1}{2}t^{-3/2}\,\frac{1}{\sqrt{4\pi}}\,e^{-\frac{x^2}{4t}} + \frac{1}{\sqrt{4\pi t}}\,e^{-\frac{x^2}{4t}}\left(-\frac{x^2}{4}\cdot\left(-\frac{1}{t^2}\right)\right) \\
&= -\frac{1}{2t\sqrt{4\pi t}}\,e^{-\frac{x^2}{4t}} + \frac{x^2}{4t^2}\cdot\frac{1}{\sqrt{4\pi t}}\,e^{-\frac{x^2}{4t}}; \qquad (1.24)\\
\frac{\partial u}{\partial x} &= -\frac{x}{2t}\cdot\frac{1}{\sqrt{4\pi t}}\,e^{-\frac{x^2}{4t}}; \\
\frac{\partial^2 u}{\partial x^2} &= -\frac{1}{2t\sqrt{4\pi t}}\,e^{-\frac{x^2}{4t}} + \frac{x^2}{4t^2}\cdot\frac{1}{\sqrt{4\pi t}}\,e^{-\frac{x^2}{4t}} \qquad (1.25)
\end{aligned}$$

From (1.24) and (1.25), we conclude that

$$\frac{\partial^2 u}{\partial x^2} = \frac{\partial u}{\partial t}. \quad \square$$

Problem 17: Show that the values of a plain vanilla put option and of a plain vanilla call option with the same maturity and strike, and on the same underlying asset, are equal if and only if the strike is equal to the forward price.

Solution: Recall that the forward price is $F = Se^{-(r-q)T}$.

From the Put–Call parity, we know that

$$C - P = Se^{-qT} - Ke^{-rT}. \qquad (1.26)$$

If a call and a put with the same strike K have the same value, i.e., if $C = P$ in (1.26), then $Se^{-qT} = Ke^{-rT}$. Thus,

$$K = Se^{(r-q)T},$$

i.e., the strike of the options is equal to the forward price. $\square$

Problem 18: Consider a portfolio with the following positions:
• long one call option with strike $K_1 = 30$;
• short two call options with strike $K_2 = 35$;
• long one call option with strike $K_3 = 40$.

All options are on the same underlying asset and have maturity T. Draw the payoff diagram at maturity of the portfolio, i.e., plot the value of the portfolio $V(T)$ at maturity as a function of $S(T)$, the price of the underlying asset at time T.

Solution: A butterfly spread is an options portfolio made of a long position in one call option with strike K_1, a long position in a call option with strike

K_3, and a short position in two calls with strike equal to the average of the strikes K_1 and K_3, i.e., with strike $K_2 = \frac{K_1+K_3}{2}$; all options have the same maturity and have the same underlying asset.

The payoff at maturity of a butterfly spread is always nonnegative, and it is positive if the price of the underlying asset at maturity is between the strikes K_1 and K_3, i.e., if $K_1 < S(T) < K_3$.

For our problem, the values of the options at maturity are, respectively,

$$\begin{aligned} C_1(T) &= \max(S(T) - K_1, 0) = \max(S(T) - 30, 0); \\ C_2(T) &= \max(S(T) - K_2, 0) = \max(S(T) - 35, 0); \\ C_3(T) &= \max(S(T) - K_3, 0) = \max(S(T) - 40, 0), \end{aligned}$$

and the value of the portfolio at maturity is

$$V(T) = C_1(T) - 2C_2(T) + C_3(T).$$

Depending on the values of the spot $S(T)$ of the underlying asset at maturity, the value $V(T)$ of the portfolio at time T is given below:

	$S(T) < 30$	$30 < S(T) < 35$	$35 < S(T) < 40$	$40 < S(T)$
$C_1(T)$	0	$S(T) - 30$	$S(T) - 30$	$S(T) - 30$
$C_2(T)$	0	0	$S(T) - 35$	$S(T) - 35$
$C_3(T)$	0	0	0	$S(T) - 40$
$V(T)$	0	$S(T) - 30$	$40 - S(T)$	0

Problem 19: Draw the payoff diagram at maturity of a bull spread with a long position in a call with strike 30 and short a call with strike 35, and of a bear spread with long a put of strike 20 and short a put of strike 15.

Solution: The payoff of the bull spread at maturity T is

$$V_1(T) = \max(S(T) - 30, 0) - \max(S(T) - 35, 0).$$

Depending on the value of the spot price $S(T)$, the value of the bull spread at maturity T is

	$S(T) < 30$	$30 < S(T) < 35$	$35 < S(T)$
$V_1(T)$	0	$S(T) - 30$	5

The value of the bear spread at maturity T is

$$V_2(T) = \max(20 - S(T), 0) - \max(15 - S(T), 0),$$

which can be written in terms of the value of $S(T)$ as

	$S(T) < 15$	$15 < S(T) < 20$	$20 < S(T)$
$V_2(T)$	5	$20 - S(T)$	0

A trader takes a long position in a bull spread if the underlying asset is expected to appreciate in value, and takes a long position in a bear spread if the value of the underlying asset is expected to depreciate. □

Problem 20: The prices of three call options with strikes 45, 50, and 55, on the same underlying asset and with the same maturity, are \$4, \$6, and \$9, respectively. Create a butterfly spread by going long a 45–call and a 55–call, and shorting two 50–calls. What are the payoff and the P&L at maturity of the butterfly spread? When would the butterfly spread be profitable? Assume, for simplicity, that interest rates are zero.

Solution: The payoff $V(T)$ of the butterfly spread at maturity is

$$V(T) = \begin{cases} 0, & \text{if } S(T) \leq 45; \\ S(T) - 45, & \text{if } 45 < S(T) \leq 50; \\ 55 - S(T), & \text{if } 50 < S(T) < 55; \\ 0, & \text{if } 55 \leq S(T). \end{cases}$$

The cost to set up the butterfly spread is

$$\$4 - \$12 + \$9 = \$1.$$

The P&L at maturity is equal to the payoff $V(T)$ minus the future value at time T of \$1, the setup cost. Since interest rates are zero, the future value of \$1 is \$1, and we conclude that

$$P\&L(T) = \begin{cases} -1, & \text{if } S(T) \leq 45; \\ S(T) - 46, & \text{if } 45 < S(T) \leq 50; \\ 54 - S(T), & \text{if } 50 < S(T) < 55; \\ -1, & \text{if } 55 \leq S(T). \end{cases}$$

The butterfly spread will be profitable if $46 < S(T) < 54$, i.e., if the spot price at maturity of the underlying asset will be between \$46 and \$54.

If $r \neq 0$, it follows similarly that the butterfly spread is profitable if

$$45 + e^{rT} < S(T) < 55 - e^{rT}. \quad \square$$

Problem 21: Which of the following two portfolios would you rather hold:

• Portfolio 1: Long one call option with strike $K = X - 5$ and long one call option with strike $K = X + 5$;

- Portfolio 2: Long two call options with strike $K = X$?
  (All options are on the same asset and have the same maturity.)

Solution: Note that being long Portfolio 1 and short Portfolio 2 is equivalent to being long a butterfly spread, and therefore will always have positive (or rather nonnegative) payoff at maturity. Therefore, if you are to *assume* a position in either one of the portfolios (not to purchase the portfolios), you are better off owning Portfolio 1, since its payoff at maturity will always be at least as big as the payoff of Portfolio 2.

To see this rigorously, denote by $V_1(t)$ and $V_2(t)$ the value of Portfolio 1 and of Portfolio 2, respectively. Let $V(t) = V_1(t) - V_2(t)$ be the value of a long position in Portfolio 1 and a short position in Portfolio 2.

The value $V(T)$ of the portfolio at the maturity T of the options is given by

$$\begin{aligned} V(T) &= V_1(T) - V_2(T) \\ &= \max(S(T) - (X-5), 0) + \max(S(T) - (X+5), 0) \\ &\quad - 2\max(S(T) - X, 0). \end{aligned}$$

It is easy to see that $V(T) \geq 0$ for any value of $S(T)$, since $V(T)$ in terms of $S(T)$ is given in the table below:

	$V(T)$
$S(T) < X - 5$	0
$X - 5 < S(T) < X$	$S(T) - (X - 5)$
$X < S(T) < X + 5$	$(X + 5) - S(T)$
$X + 5 < S(T)$	0

Problem 22: A stock with spot price \$42 pays dividends continuously at a rate of 3%. The four months put and call options with strike 40 on this asset are trading at \$2 and \$4, respectively. The risk-free rate is constant and equal to 5% for all times. Show that the Put-Call parity is not satisfied and explain how would you take advantage of this arbitrage opportunity.

Solution: The following values are given: $S = 42$; $K = 40$; $T = 1/3$; $r = 0.05$; $q = 0.03$; $P = 2$; $C = 4$.

The Put–Call parity is not satisfied, since

$$P + Se^{-qT} - C = 39.5821 > 39.3389 = Ke^{-rT}. \qquad (1.27)$$

Therefore, a riskless profit can be obtained by "buying low and selling high", i.e., by selling the portfolio on the left hand side of (1.27) and buying the portfolio on the right hand side of (1.27) (which is cash only). The riskless profit at maturity will be the future value at time T of the mispricing from the Put–Call parity, i.e.,

$$(39.5821 - 39.3389)e^{rT} \;=\; 0.2473. \tag{1.28}$$

To show this, start with no money and sell one put option, short e^{-qT} shares, and buy one call option. This will generate the following cash amount:

$$P + Se^{-qT} - C \;=\; 39.5821,$$

since shorting the shares means that e^{-qT} shares are borrowed and sold on the market for cash. (The short will be closed at maturity T by buying shares on the market and returning them to the borrower; see below for more details.)

At time 0, the portfolio consists of the following positions:
- short one put option with strike K and maturity T;
- short e^{-qT} shares;
- long one call option with strike K and maturity T;
- cash: +$39.5821.

The initial value of the portfolio is zero, since no money were invested:

$$V(0) \;=\; -P(0) - S(0)e^{-qT} + C(0) + 39.5821 \;=\; 0.$$

Note that by shorting the shares you are responsible for paying the accrued dividends. Assume that the dividend payments are financed by shorting more shares of the underlying asset and using the cash proceeds to make the dividend payments. Then, the short position in e^{-qT} shares at time 0 will become a short position in one share[3] at time T.

The value of the portfolio at maturity is

$$V(T) \;=\; -P(T) - S(T) + C(T) + 39.5821e^{rT}.$$

As shown when proving the Put–Call parity,

$$P(T)+S(T)-C(T) \;=\; \max(K-S(T),0)+S(T)-\max(S(T)-K,0) \;=\; K,$$

regardless of the value $S(T)$ of the underlying asset at maturity.

Therefore,

$$\begin{aligned} V(T) &= -(P(T) + S(T) - C(T)) + 39.5821e^{rT} \\ &= -K + 39.5821e^{rT} \;=\; -40 + 40.2473 \;=\; 0.2473. \end{aligned}$$

This value represents the risk–free profit made by exploiting the discrepancy from the Put–Call parity, and is the same as the future value at time T of the mispricing from the Put–Call parity; cf. (1.28). □

[3] This is similar to converting a long position in e^{-qT} shares at time 0 into a long position in one share at time T, through continuous purchases of (fractions of) shares using the dividend payments, which is a more intuitive process.

Problem 23: The bid and ask prices for a six months European call option with strike 40 on a non–dividend–paying stock with spot price 42 are \$5 and \$5.5, respectively. The bid and ask prices for a six months European put option with strike 40 on the same underlying asset are \$2.75 and \$3.25, respectively. Assume that the risk free rate is equal to 0. Is there an arbitrage opportunity present?

Solution: For $r = 0$, the Put–Call parity becomes $P + S - C = K$, which in this case can be written as $C - P = 2$.

Thus, an arbitrage occurs if $C - P$ can be "bought" for less than \$2 (i.e., if a call option is bought and a put option is sold for less than \$2), or if $C - P$ can be "sold" for more than \$2 (i.e., if a call option can be sold and a put option can be bought for more than \$2).

From the bid and ask prices, we find that the call can be bought for \$5.5 and the put can be sold for \$2.75. Then, $C - P$ can be "bought" for \$5.5-\$2.75=\$2.75, which is more than \$2. Therefore, no risk-free profit can be achieved this way.

Also, a call can be sold for \$5 and a put can be bought for \$3.25. Therefore, $C - P$ can be "sold" for \$5-\$3.25=\$1.75, which is less than \$2. Again, no risk-free profit can be achieved. □

Problem 24: Denote by C_{bid} and C_{ask}, and by P_{bid} and P_{ask}, respectively, the bid and ask prices for a plain vanilla European call and for a plain vanilla European put option, both with the same strike K and maturity T, and on the same underlying asset with spot price S and paying dividends continuously at rate q. Assume that the risk–free interest rates are constant equal to r.

Find necessary and sufficient no–arbitrage conditions for C_{bid}, C_{ask}, P_{bid}, and P_{ask}.

Solution: Recall the Put–Call parity

$$C - P \;=\; Se^{-qT} - Ke^{-rT},$$

where the right hand represents the value of a forward contract on the underlying asset with strike K.

An arbitrage would exist in one of the following two instances:

• if the purchase price of a long call short put portfolio, i.e., $C_{ask} - P_{bid}$ were less than the value $Se^{-qT} - Ke^{-rT}$ of the forward contract, i.e., if

$$C_{ask} - P_{bid} \;<\; Se^{-qT} - Ke^{-rT};$$

• if the selling price of a long call short put portfolio, i.e., $C_{bid} - P_{ask}$ were greater than the value $Se^{-qT} - Ke^{-rT}$ of the forward contract, i.e., if

$$C_{bid} - P_{ask} \;>\; Se^{-qT} - Ke^{-rT}.$$

We conclude that there is no–arbitrage directly following from the Put–Call parity if and only if

$$C_{bid} - P_{ask} \leq Se^{-qT} - Ke^{-rT} \leq C_{ask} - P_{bid}. \tag{1.29}$$

Note that the no–arbitrage condition (1.29) can also be written as

$$C_{bid} - P_{ask} \leq F \leq C_{ask} - P_{bid},$$

where $F = Se^{-qT} - Ke^{-rT}$ is the value of a forward contract with delivery price K and maturity T on the same underlying asset. □

Problem 25: You expect that an asset with spot price \$35 will trade in the \$40–\$45 range in one year. One year at–the–money calls on the asset can be bought for \$4. To act on the expected stock price appreciation, you decide to either buy the asset, or to buy ATM calls. Which strategy is better, depending on where the asset price will be in a year?

Solution: For every \$1000 invested, the payoff in one year of the first strategy, i.e., of buying the asset, is

$$V_1(T) = \frac{1000}{35} S(T),$$

where $S(T)$ is the spot price of the asset in one year.

For every \$1000 invested, the payoff in one year of the second strategy, i.e., of investing everything in buying call options, is

$$\begin{aligned} V_2(T) &= \frac{1000}{4} \max(S(T) - 35, 0) \\ &= \begin{cases} \frac{1000}{4}(S(T) - 35), & \text{if } S(T) \geq 35; \\ 0, & \text{if } S(T) < 35. \end{cases} \end{aligned}$$

If $S(T)$ is less than \$35, the calls expire worthless and the speculative strategy of investing everything in call options will lose all the money invested in it, while the first strategy of buying the asset will not lose all its value. However, investing everything in the call options is very profitable if the asset appreciates in value, i.e., is $S(T)$ is significantly larger than \$35. The breakeven point of the two strategies, i.e., the spot price at maturity of the underlying asset where both strategies have the same payoff is \$39.5161, since

$$\frac{1000}{35} S(T) = \frac{1000}{4}(S(T) - 35) \Longleftrightarrow S(T) = 39.5161.$$

If the price of the asset will, indeed, be in the \$40–\$45 range in one year, then buying the call options is the more profitable strategy. □

Problem 26: Create a portfolio with the following payoff at time T:

$$V(T) = \begin{cases} 2S(T), & \text{if } S(T) < 20; \\ 60 - S(T), & \text{if } 20 \leq S(T) < 40; \\ S(T) - 20, & \text{if } 40 \leq S(T), \end{cases} \tag{1.30}$$

where $S(T)$ is the spot price at time T of a given asset. Use plain vanilla options with maturity T as well as cash positions and positions in the asset itself. Assume, for simplicity, that the asset does not pay dividends and that interest rates are zero.

Solution: Using plain vanilla options, cash, and the underlying asset the payoff $V(T)$ can be replicated in different ways.

One way is to use the underlying asset, calls with strike 20, and calls with strike 40.

First of all, a portfolio with a long position in two units of the underlying asset has value $2S(T)$ at maturity, when $S(T) < 20$.

To replicate the portfolio payoff $60 - S(T)$ when $20 \leq S(T) < 40$, note that

$$60 - S(T) = 2S(T) + 60 - 3S(T) = 2S(T) - 3(S(T) - 20).$$

This is equivalent to a long position in two units of the underlying asset and a short position in three calls with strike 20.

To replicate the portfolio payoff $S(T) - 20$ when $40 \leq S(T)$, note that

$$S(T) - 20 = 60 - S(T) + 2S(T) - 80 = 2S(T) - 3(S(T) - 20) + 2(S(T) - 40).$$

This is equivalent to a long position in two units of the underlying asset, a short position in three calls with strike 20, and a long position in two calls with strike 40.

Summarizing, the replicating portfolio is made of

- long two units of the asset;
- short 3 call options with strike $K = 20$ on the asset;
- long 2 call options with strike $K = 40$ on the asset.

We check that the payoff of this portfolio at maturity, i.e.,

$$V_1(T) = 2S(T) - 3\max(S(T) - 20, 0) + 2\max(S(T) - 40, 0) \tag{1.31}$$

is the same as the payoff from (1.30):

	$V_1(T)$
$S(T) < 20$	$2S(T)$
$20 \leq S(T) < 40$	$2S(T) - 3(S(T) - 20) = 60 - S(T)$
$40 \leq S(T)$	$60 - S(T) + 2(S(T) - 40) = S(T) - 20$

As a consequence of the Put–Call parity, it follows that the payoff $V(T)$ from (1.30) can also be synthesized using put options. If the asset does not pay dividends and if interest rates are zero, then, from the Put–Call parity, it follows that

$$C = P + S - K.$$

Denote by C_{20} and P_{20}, and by C_{40} and P_{40}, the values of the call and put options with strikes 20 and 40, respectively.

Then, the replicating portfolio with payoff at maturity given by (1.31) can be written as

$$V = 2S - 3C_{20} + 2C_{40}. \tag{1.32}$$

To synthesize a short position in three calls with strike 20, note that

$$-3C_{20} = -3P_{20} - 3S + 60, \tag{1.33}$$

which is equivalent to taking a short position in three units of the underlying asset, taking a short position in three put options with strike 20, and being a long \$60.

Similarly, to synthesize a long position in two calls with strike 40, note that

$$2C_{40} = 2P_{40} + 2S - 80, \tag{1.34}$$

which is equivalent to a borrowing \$80, taking a long position in two units of the underlying asset, and taking a long position in two put options with strike 40.

Using (1.33) and (1.34), we obtain that the payoff at maturity given by (1.31) can be replicated using the following portfolio consisting of put options, cash, and the underlying asset:

$$\begin{aligned} V &= 2S - 3C_{20} + 2C_{40} \\ &= 2S - 3P_{20} - 3S + 60 + 2P_{40} + 2S - 80 \\ &= S - 3P_{20} + 2P_{40} - 20. \end{aligned} \tag{1.35}$$

The positions of the replicating portfolio (1.35) can be summarized as follows:

- long one unit of the asset;
- short \$20 cash;
- short 3 put options with strike $K = 20$ on the asset;
- long 2 put options with strike $K = 40$ on the asset.

We check that the payoff of this portfolio at maturity, i.e.,

$$V_2(T) = S(T) - 20 - 3\max(20 - S(T), 0) + 2\max(40 - S(T), 0)$$

is the same as the payoff from (1.30):

	$V_1(T)$
$S(T) \leq 20$	$S(T) - 20 - 3(20 - S(T)) + 2(40 - S(T)) = 2S(T)$
$20 < S(T) \leq 40$	$S(T) - 20 + 2(40 - S(T)) = 60 - S(T)$
$40 < S(T)$	$S(T) - 20$

If the asset pays dividends continuously at rate q and if interest rates are constant and equal to r, in order to obtain the same payoffs at maturity, the asset positions in the two portfolios must be adjusted as follows:

The first replicating portfolio will be made of the following positions:
- long $2e^{-qT}$ units of the asset;
- short 3 call options with strike $K = 20$ on the asset;
- long 2 call options with strike $K = 40$ on the asset.

The second replicating portfolio will be made of the following positions:
- long e^{-qT} units of the asset;
- short $\$20e^{-rT}$ cash;
- short 3 put options with strike $K = 20$ on the asset;
- long 2 put options with strike $K = 40$ on the asset.

Note that any piecewise linear payoff of a single asset can be synthesized, in theory, by using plain vanilla options, cash and asset positions. □

Problem 27: A derivative security pays a cash amount c if the spot price of the underlying asset at maturity is between K_1 and K_2, where $0 < K_1 < K_2$, and expires worthless otherwise. How do you synthesize this derivative security (i.e., how do you recreate its payoff almost exactly) using plain vanilla call options?

Solution: The payoff of the derivative security is

$$V(T) = \begin{cases} 0, & \text{if } S(T) \leq K_1; \\ c, & \text{if } K_1 < S(T) < K_2; \\ 0, & \text{if } K_2 \leq S(T). \end{cases}$$

Since $V(T)$ is discontinuous, it cannot be replicated exactly using call options, whose payoffs are continuous.

We approximate the payoff $V(T)$ of the derivative security by the following payoff

$$V_\epsilon(T) = \begin{cases} 0, & \text{if } S(T) < K_1 - \epsilon; \\ \frac{c}{\epsilon}(S(T) - (K_1 - \epsilon)), & \text{if } K_1 - \epsilon \le S(T) \le K_1; \\ c, & \text{if } K_1 < S(T) < K_2; \\ c - \frac{c}{\epsilon}(S(T) - K_2), & \text{if } K_2 \le S(T) \le K_2 + \epsilon; \\ 0, & \text{if } K_2 + \epsilon < S(T). \end{cases} \quad (1.36)$$

Note that $V(T) = V_\epsilon(T)$ unless the value $S(T)$ of the underlying asset at maturity is either between $K_1 - \epsilon$ and K_1, or between K_2 and $K_2 + \epsilon$.

The payoff $V_\epsilon(T)$ can be realized by going long c/ϵ bull spreads with strikes $K_1 - \epsilon$ and K_1, and shorting c/ϵ bull spreads with strikes K_2 and $K_2 + \epsilon$.

The payoff $V(T)$ of the given derivative security can be synthesized by taking the following positions:

- long c/ϵ calls with strike $K_1 - \epsilon$;
- short c/ϵ calls with strike K_1;
- short c/ϵ calls with strike K_2;
- long c/ϵ calls with strike $K_2 + \epsilon$.

It is easy to see that the payoff $V_\epsilon(T)$ is the same as in (1.36):

	$V_\epsilon(T)$
$S(T) < K_1 - \epsilon$	0
$K_1 - \epsilon \le S(T) < K_1$	$\frac{c}{\epsilon}(S(T) - (K_1 - \epsilon))$
$K_1 \le S(T) < K_2$	$\frac{c}{\epsilon}(S(T) - (K_1 - \epsilon)) - \frac{c}{\epsilon}(S(T) - K_1)) = c$
$K_2 \le S(T) < K_2 + \epsilon$	$c - \frac{c}{\epsilon}(S(T) - K_2)$
$K_2 + \epsilon < S(T)$	$c - \frac{c}{\epsilon}(S(T) - K_2) + \frac{c}{\epsilon}(S(T) - (K_2 + \epsilon)) = 0$

Problem 28: Call options with strikes 100, 120, and 130 on the same underlying asset and with the same maturity are trading for 8, 5, and 3, respectively (there is no bid–ask spread). Is there an arbitrage opportunity present? If yes, how can you make a riskless profit?

Note: A solution to this problem based on the convexity of the payoff of call and put options is discussed in section 1.3 at the end of this chapter.

Solution: For an arbitrage opportunity to be present, there must be a portfolio made of the three options with nonnegative payoff at maturity and with a negative cost of setting up.

Let $K_1 = 100 < K_2 = 120 < K_3 = 130$ be the strikes of the options. Denote by x_1, x_2, x_3 the options positions (which can be either negative or positive) at time 0. Then, at time 0, the portfolio is worth

$$V(0) = x_1C_1(0) + x_2C_2(0) + x_3C_3(0)$$

At maturity T, the value of the portfolio will be

$$\begin{aligned} V(T) &= x_1C_1(T) + x_2C_2(T) + x_3C_3(T) \\ &= x_1\max(S(T)-K_1,0) + x_2\max(S(T)-K_2,0) \\ &\quad + x_3\max(S(T)-K_3,0), \end{aligned}$$

respectively.

Depending on the value $S(T)$ of the underlying asset at maturity, the value $V(T)$ of the portfolio is as follows:

	$V(T)$
$S(T) < K_1$	0
$K_1 < S(T) < K_2$	$x_1S(T) - x_1K_1$
$K_2 < S(T) < K_3$	$(x_1+x_2)S(T) - x_1K_1 - x_2K_2$
$K_3 < S(T)$	$(x_1+x_2+x_3)S(T) - x_1K_1 - x_2K_2 - x_3K_3$

Note that $V(T)$ is nonnegative when $S(T) \leq K_2$ only if a long position is taken in the option with strike K_1, i.e., if $x_1 \geq 0$. The payoff $V(T)$ decreases when $K_2 < S(T) < K_3$, accounting for the short position in the two call options with strike K_2, and then increases when $S(T) \geq K_3$.

We conclude that $V(T) \geq 0$ for any value of S(T) if and only if $x_1 \geq 0$, if the value of the portfolio when $S(T) = K_3$ is nonnegative, i.e., if $(x_1 + x_2)K_3 - x_1K_1 - x_2K_2 \geq 0$, and if $x_1 + x_2 + x_3 \geq 0$.

Thus, an arbitrage exists if and only if the values $C_1(0)$, $C_2(0)$, $C_3(0)$ are such that we can find x_1, x_2, and x_3 with the following properties:

$$\begin{aligned} x_1C_1(0) + x_2C_2(0) + x_3C_3(0) &< 0; \\ x_1 &\geq 0; \\ (x_1+x_2)K_3 - x_1K_1 - x_2K_2 &\geq 0; \\ x_1 + x_2 + x_3 &\geq 0. \end{aligned}$$

For $C_1(0) = 8$, $C_2(0) = 5$, $C_3(0) = 3$ and $K_1 = 100$, $K_2 = 120$, $K_3 = 130$, the problem becomes finding $x_1 \geq 0$, and x_2 and x_3 such that

$$8x_1 + 5x_2 + 3x_3 < 0; \quad (1.37)$$

$$30x_1 + 10x_2 \geq 0; \quad (1.38)$$

$$x_1 + x_2 + x_3 \geq 0. \quad (1.39)$$

(For these option prices, arbitrage will be possible since the middle option is overpriced relative to the other two options.)

The easiest way to find values of x_1, x_2, and x_3 satisfying the constraints above is to note that arbitrage can occur for a portfolio with long positions in the options with lowest and highest strikes, and with a short position in the option with middle strike (note the similarity to butterfly spreads). Then, choosing $x_3 = -x_1 - x_2$ would be optimal; cf. (1.39). The constraints (1.37) and (1.38) become

$$\begin{aligned} 5x_1 + 2x_2 &< 0; \\ 3x_1 + x_2 &\geq 0. \end{aligned}$$

These constraints are satisfied, e.g., for $x_1 = 1$ and $x_2 = -3$, which corresponds to $x_3 = 2$.

Buying one option with strike 100, selling three options with strike 120, and buying two options with strike 130 will generate a positive cash flow of \$1, and will result in a portfolio that will not lose money, regardless of the value of the underlying asset at the maturity of the options. □

Problem 29: Call options on the same underlying asset and with the same maturity, with strikes $K_1 < K_2 < K_3$, are trading for C_1, C_2 and C_3, respectively (no Bid–Ask spread), with $C_1 > C_2 > C_3$. Find necessary and sufficient conditions on the prices C_1, C_2 and C_3 such that no–arbitrage exists corresponding to a portfolio made of positions in the three options.

Solution: An arbitrage exists if and only if a no–cost portfolio can be set up with non–negative payoff at maturity regardless of the price of the underlying asset at maturity, and such that the probability of a strictly positive payoff is greater than 0.

Consider a portfolio made of positions in the three options with value 0 at inception, and let $x_i > 0$ be the size of the portfolio position in the option with strike K_i, for $i = 1:3$. Let $S = S(T)$ be the value of the underlying asset at maturity. For no–arbitrage to occur, there are three possibilities:

Portfolio 1: Long the K_1–option, short the K_2–option, long the K_3–option.

Arbitrage exists if we can find $x_i > 0$, $i = 1:3$, such that

$$x_1 C_1 - x_2 C_2 + x_3 C_3 = 0; \tag{1.40}$$

$$x_1(S - K_1) - x_2(S - K_2) + x_3(S - K_3) \geq 0, \quad \forall\, S \geq 0. \tag{1.41}$$

We note that (1.41) holds if and only if the following two conditions are satisfied:

$$x_1 - x_2 + x_3 \geq 0; \tag{1.42}$$

$$x_1(K_3 - K_1) - x_2(K_3 - K_2) \geq 0. \tag{1.43}$$

We solve (1.40) for x_3 and obtain

$$x_3 = x_2\frac{C_2}{C_3} - x_1\frac{C_1}{C_3}. \qquad (1.44)$$

Since we assumed that $x_3 > 0$, the following condition must also be satisfied:

$$\frac{x_2}{x_1} > \frac{C_1}{C_2}. \qquad (1.45)$$

Recall that $C_1 > C_2 > C_3$. Using the value of x_3 from (1.44), it follows that (1.42) and (1.43) hold true if and only if

$$\frac{x_2}{x_1} \geq \frac{C_1 - C_3}{C_2 - C_3}; \qquad (1.46)$$

$$\frac{x_2}{x_1} \leq \frac{K_3 - K_1}{K_3 - K_2}. \qquad (1.47)$$

Also, note that if (1.46) holds true, then (1.45) is satisfied as well, since

$$\frac{C_1 - C_3}{C_2 - C_3} > \frac{C_1}{C_2}.$$

We conclude that arbitrage happens if and only if we can find $x_1 > 0$ and $x_2 > 0$ such that (1.46) and (1.47) are simultaneously satisfied. Therefore, no–arbitrage exists if and only if

$$\frac{K_3 - K_1}{K_3 - K_2} < \frac{C_1 - C_3}{C_2 - C_3}. \qquad (1.48)$$

Portfolio 2: Long the K_1–option, short the K_2–option, short the K_3–option.

Arbitrage exists if we can find $x_i > 0$, $i = 1 : 3$, such that

$$x_1C_1 - x_2C_2 - x_3C_3 = 0; \qquad (1.49)$$

$$x_1(S - K_1) - x_2(S - K_2) - x_3(S - K_3) \geq 0, \ \forall\, S \geq 0. \qquad (1.50)$$

The inequality (1.50) holds if and only if the following two conditions are satisfied:

$$x_1 - x_2 - x_3 \geq 0; \qquad (1.51)$$

$$x_1(K_3 - K_1) - x_2(K_3 - K_2) \geq 0. \qquad (1.52)$$

However, (1.49) and (1.51) cannot be simultaneously satisfied. Since $C_1 > C_2 > C_3$, it is easy to see that

$$x_1 = x_2\frac{C_2}{C_1} - x_3\frac{C_3}{C_1} < x_2 + x_3.$$

In other words, no arbitrage can be obtained by being long the option with strike K_1 and short the options with strikes K_2 and K_3.

Portfolio 3: Long the K_1–option, long the K_2–option, short the K_3–option.

Arbitrage exists if we can find $x_i > 0$, $i = 1 : 3$, such that

$$x_1 C_1 + x_2 C_2 - x_3 C_3 \; = \; 0; \tag{1.53}$$

$$x_1(S - K_1) + x_2(S - K_2) - x_3(S - K_3) \; \geq \; 0, \quad \forall \; S \geq 0. \tag{1.54}$$

The inequality (1.54) holds if and only if

$$x_1 + x_2 - x_3 \; \geq \; 0. \tag{1.55}$$

It is easy to see that (1.53) and (1.55) cannot be simultaneously satisfied:

$$x_3 \; = \; x_1 \frac{C_1}{C_3} + x_2 \frac{C_2}{C_3} \; > \; x_1 + x_2,$$

since $C_1 > C_2 > C_3$.

Therefore, no arbitrage can be obtained by being long the options with strikes K_1 and K_2 and short the option with strike K_3.

We conclude that (1.48), i.e.,

$$\frac{K_3 - K_1}{K_3 - K_2} \; < \; \frac{C_1 - C_3}{C_2 - C_3} \tag{1.56}$$

is the only condition required for no–arbitrage. □

Problem 30: The risk free rate is 8% compounded continuously and the dividend yield of a stock index is 3%. The index is at 12,000 and the futures price of a contract deliverable in three months is 12,100. Is there an arbitrage opportunity, and how do you take advantage of it?

Solution: The arbitrage–free futures price of the futures contract is

$$12000 e^{(r-q)T} \; = \; 12000 e^{(0.08-0.03)/4} \; = \; 12150.94 \; > \; 12100.$$

Therefore, the futures contract is underpriced and should be bought while hedged statically by shorting $e^{-qT} = 0.9925$ units of index for each futures contract that is sold.

At maturity, the asset is bought for 12100 and the short is closed (the dividends paid on the short position increase the size of the short position to 1 unit of the index). The realized gain is the interest accrued on the cash resulting from the short position minus 12100, i.e.,

$$e^{0.08/4} \left(e^{-0.03/4} \cdot 12000 \right) \; - \; 12100 \; = \; 50.94. \quad \square$$

1.3 Arbitrage and the convexity of option values

One of the problems in this chapter (see Problem 28) asks to find an arbitrage given call options with strikes 100, 120, and 130 (on the same underlying asset and with the same maturity) and corresponding prices \$8, \$5, and \$3, respectively.

The arbitrage opportunity is based on the fact that the convexity of the call options price with respect to strike is violated.

Recall that a function $f : \mathbb{R} \to \mathbb{R}$ is strictly convex (upward) if and only if[4]

$$f(\lambda x + (1-\lambda)y) \;<\; \lambda f(x) + (1-\lambda)f(y), \quad \forall\, x, y \in \mathbb{R}, \quad \forall\, \lambda \in (0,1). \tag{1.57}$$

A simple way to see that the values of plain vanilla call and put options are strictly convex functions of the strike is by contradiction using (1.57). For example, for call options, assume that there exist two strikes K_1 and K_3 with $0 < K_1 < K_3$, and a parameter $\lambda \in (0,1)$ such that

$$C(\lambda K_1 + (1-\lambda)K_3) \;\geq\; \lambda C(K_1) + (1-\lambda)C(K_3), \tag{1.58}$$

where $C(K)$ denotes the value at time 0 of a call option with strike K.

Let

$$K_2 \;=\; \lambda K_1 + (1-\lambda)K_3. \tag{1.59}$$

Since $0 < \lambda < 1$, it follows that $K_1 < K_2 < K_3$.

Consider the following portfolio:

- long λ call options with strike K_1;
- long $1-\lambda$ call options with strike K_3;
- short one call option with strike K_2;

The setup cost[5] for this portfolio is nonnegative, since

$$\begin{aligned} & C(K_2) - \lambda C(K_1) - (1-\lambda)C(K_3) \\ = \;& C(\lambda K_1 + (1-\lambda)K_3) - \lambda C(K_1) - (1-\lambda)C(K_3) \\ \geq \;& 0; \end{aligned}$$

cf. assumption (1.58).

Let $V(T)$ be the value of the portfolio at time T, the maturity of the options. It is easy to see that $V(T)$ is nonnegative regardless of the value $S(T)$ of the underlying asset at maturity:

[4]If $f''(x)$ exists and is continuous, condition (1.57) is equivalent to $f''(x) > 0$, $\forall\, x \in \mathbb{R}$.

[5]Note that the set up cost for a long position with value V_0 is -\$$V_0$, since the amount \$$V_0$ is paid in order to establish the long position. Similarly, the set up cost for a short position with value V_0 is \$$V_0$, since the amount \$$V_0$ is received when establishing the short position.

(i) If $S(T) \leq K_1$, then $V(T) = 0$;
(ii) If $K_1 < S(T) \leq K_2$, then

$$V(T) = \lambda(S(T) - K_1) > 0.$$

(iii) If $K_2 < S(T) \leq K_3$, then

$$\begin{aligned} V(T) &= \lambda(S(T) - K_1) - (S(T) - K_2) \\ &= K_2 - \lambda K_1 - (1 - \lambda)S(T) \\ &= (\lambda K_1 + (1 - \lambda)K_3) - \lambda K_1 - (1 - \lambda)S(T) \\ &= (1 - \lambda)(K_3 - S(T)) \\ &\geq 0, \end{aligned}$$

where we used the fact that $K_2 = \lambda K_1 + (1 - \lambda)K_3$; cf. (1.59).
(iv) If $K_3 < S(T)$, then

$$\begin{aligned} V(T) &= \lambda(S(T) - K_1) - (S(T) - K_2) + (1 - \lambda)(S(T) - K_3) \\ &= K_2 - \lambda K_1 - (1 - \lambda)K_3 \\ &= 0; \end{aligned}$$

cf. (1.59).

The portfolio considered above generates cash upon setting up and has nonnegative payoff at options maturity regardless of the state of the market and has positive payoffs for certain market states. According to the Generalized Law of One Price (Theorem 1.10 in [1]), this is an arbitrage opportunity. This is due to the fact that the option prices do not satisfy the requirement of convexity with respect to strike prices; see (1.58). Thus, in arbitrage–free markets, the values of plain–vanilla call options must be strictly convex functions of their strikes.

Following a similar reasoning by contradiction, we can show that the values of plain–vanilla put options are strictly convex functions of the strike price.

The convexity of options values can be used to find an arbitrage for the options in Problem 28. We first show that if otherwise identical call options with strikes 100, 120, and 130 have corresponding prices \$8, \$5, and \$3, respectively, then the convexity of the option values is violated.

In the $(K, C(K))$ space, the equation of the straight line passing through the points $(100, 8)$ and $(130, 3)$ is

$$y = \frac{8(130 - x)}{30} + \frac{3(x - 100)}{30}, \tag{1.60}$$

where the variable x corresponds to K, and the variable y corresponds to $C(K)$. The point on this line corresponding to strike 120 is obtained by substituting $x = 120$ in (1.60), and is equal to

$$8 \cdot \frac{1}{3} + 3 \cdot \frac{2}{3} = \frac{14}{3}.$$

Since $C(K)$ is a strictly convex function of K, the value of the call option with strike 120 should be below the straight line passing through the price points of the options with strikes 100 and 130; cf. (1.57). However, $C(120) = 5 > \frac{14}{3}$, and the convexity of the call option value as function of strike is not satisfied.

Thus, an arbitrage exists. Using a "buy low, sell high" strategy, we could buy $\frac{1}{3}$ options with strike 100, buy $\frac{2}{3}$ options with strike 130, and sell 1 option with strike 120. To avoid fractions, we set up the following portfolio:

- long 1 call with strike 100;
- long 2 calls with strike 130;
- short 3 calls with strike 120.

There is a $1 positive cash flow at the set up of this portfolio, since

$$3 \cdot \$5 \; - \; \$8 \; - \; 2 \cdot \$3 \; = \; \$1.$$

At the maturity of the options, the value $V(T)$ of the portfolio will be nonnegative:

	$V(T)$
$S(T) \leq 100$	0
$100 < S(T) \leq 120$	$S(T) - 100 > 0$
$120 < S(T) \leq 130$	$(S(T) - 100) - 3(S(T) - 120) \; = \; 260 - 2S(T) \geq 0$
$130 < S(T)$	$260 - 2S(T) + 2(S(T) - 130) \; = \; 0$

The future value at time T of the $1 cash flow from the set up cost of the portfolio is therefore risk–free profit.

A similar arbitrage opportunity due to the lack of convexity of put option prices can be found in the example below:

Example: The prices of three put options with strikes 40, 50, and 70, but otherwise identical, are $10, $20, and $30, respectively. Find an arbitrage opportunity.

Answer: In the $(K, P(K))$ space, the equation of the straight line passing through the points $(40, 10)$ and $(70, 30)$ is

$$y \;=\; \frac{10(70 - x)}{30} + \frac{30(x - 40)}{30}, \tag{1.61}$$

where the variable x corresponds to K, and the variable y corresponds to $P(K)$. The point on this line corresponding to strike 50 is obtained by substituting $x = 50$ in (1.61), and is equal to

$$10 \cdot \frac{2}{3} + 30 \cdot \frac{1}{3} = \frac{50}{3}.$$

Since $P(K)$ is a strictly convex function of K, the value of the put option with strike 50 should be below the straight line passing through the price points of the options with strikes 40 and 70. However, $P(50) = 20 > \frac{50}{3}$, and the convexity of the put option value as function of strike is not satisfied.

Thus, an arbitrage exists. Using a "buy low, sell high" strategy, we could buy $\frac{2}{3}$ options with strike 40, buy $\frac{1}{3}$ options with strike 70, and sell 1 option with strike 50. To avoid fractions, we set up the following portfolio:

- long 2 put options with strike 40;
- long 1 put option with strike 70;
- short 3 put options with strike 50.

There is a \$10 positive cash flow at the set up of this portfolio, since

$$3 \cdot \$20 \ - \ 2 \cdot \$10 \ - \ \$30 \ = \ \$10.$$

At the maturity of the options, the value $V(T)$ of the portfolio will be nonnegative:

	$V(T)$
$70 \leq S(T)$	0
$50 \leq S(T) < 70$	$70 - S(T)$
$40 \leq S(T) < 50$	$(70 - S(T)) - 3(50 - S(T)) \ = \ 2S(T) - 80 \geq 0$
$S(T) < 40$	$2S(T) - 80 + 2(40 - S(T)) \ = \ 0$

The future value at time T of the \$10 cash flow from the set up cost of the portfolio is therefore risk–free profit. □

We conclude by giving further insight into the general no–arbitrage condition (1.56), i.e.,

$$\frac{K_3 - K_1}{K_3 - K_2} < \frac{C_1 - C_3}{C_2 - C_3}, \tag{1.62}$$

derived in Problem 29 as the necessary and sufficient no arbitrage condition for three call options with prices C_1, C_2, C_3 and strikes K_1, K_2, K_3, with $K_1 < K_2 < K_3$.

In the $(K, C(K))$ space, the equation of the straight line passing through the points (K_1, C_1) and (K_3, C_3) is

$$y \ = \ \frac{C_1(K_3 - x)}{K_3 - K_1} \ + \ \frac{C_3(x - K_1)}{K_3 - K_1}, \tag{1.63}$$

where the variable x corresponds to K, and the variable y corresponds to $C(K)$. The point on this line corresponding to strike K_2 is obtained by substituting $x = K_2$ in (1.63), and is equal to

$$\frac{C_1(K_3 - K_2)}{K_3 - K_1} + \frac{C_3(K_2 - K_1)}{K_3 - K_1}.$$

If

$$C_2 \geq \frac{C_1(K_3 - K_2)}{K_3 - K_1} + \frac{C_3(K_2 - K_1)}{K_3 - K_1},$$

then the value of the call option with strike K_2 would not be below the straight line passing through the price points of the options with strikes K_1 and K_3. An arbitrage opportunity would then be present, since the convexity of the option values as functions of strikes would be violated. A portfolio consisting of a long position of size $\frac{K_3-K_2}{K_3-K_1}$ in the option with strike K_1, a short position in one option with strike K_2, and a long position of size $\frac{K_2-K_1}{K_3-K_1}$ in the option with strike K_3 would have a nonnegative setup cost and a nonnegative payoff at options maturity, with a strictly positive payoff if $K_1 < S(T) < K_3$.

Thus, a necessary no–arbitrage condition is

$$C_2 < \frac{C_1(K_3 - K_2)}{K_3 - K_1} + \frac{C_3(K_2 - K_1)}{K_3 - K_1}. \tag{1.64}$$

It is easy to see that (1.62) and (1.64) are both equivalent to

$$C_2(K_3 - K_1) < C_1(K_3 - K_2) + C_3(K_2 - K_1),$$

and therefore are the same.

Note that the proof of Problem 29 shows that condition (1.64) is also sufficient.

Chapter 2

Improper integrals. Numerical integration. Interest rates. Bonds.

2.1 Exercises

1. Compute

$$\lim_{x \to 0} \frac{N(x) - \frac{1}{2} - \frac{x}{\sqrt{2\pi}}}{x^3},$$

where

$$N(x) = \frac{1}{\sqrt{2\pi}} \int_{-\infty}^{x} e^{-\frac{y^2}{2}} \, dy$$

is the cumulative density function of the standard normal variable.

2. Let $f : (0, \infty) \to \mathbb{R}$ denote the Gamma function, i.e., let

$$f(\alpha) = \int_0^\infty x^{\alpha - 1} e^{-x} \, dx.$$

(i) Show that $f(\alpha)$ is well defined for any $\alpha > 0$, i.e., show that both

$$\int_0^1 x^{\alpha-1} e^{-x} \, dx = \lim_{t \searrow 0} \int_t^1 x^{\alpha-1} e^{-x} \, dx$$

and

$$\int_1^\infty x^{\alpha-1} e^{-x} \, dx = \lim_{t \to \infty} \int_1^t x^{\alpha-1} e^{-x} \, dx$$

exist and are finite.

(ii) Prove, using integration by parts, that $f(\alpha) = (\alpha - 1) \, f(\alpha - 1)$ for any $\alpha > 1$. Show that $f(1) = 1$ and conclude that, for any $n \geq 1$ positive integer, $f(n) = (n-1)!$.

3. Compute an approximate value of

$$\int_1^3 \sqrt{x}\, e^{-x} dx$$

using the Midpoint rule, the Trapezoidal rule, and Simpson's rule. Start with $n = 4$ intervals, and double the number of intervals until two consecutive approximations are within 10^{-6} of each other.

4. Let $f : \mathbb{R} \to \mathbb{R}$ given by

$$f(x) = \frac{x^{5/2}}{1+x^2}.$$

(i) Use Midpoint rule with $tol = 10^{-6}$ to compute an approximation of

$$I = \int_0^1 f(x)\, dx = \int_0^1 \frac{x^{5/2}}{1+x^2}. \tag{2.1}$$

(ii) Show that $f^{(4)}(x)$ is not bounded on the interval $(0, 1)$.

(iii) Apply Simpson's rule with $n = 2^k$, $k = 2 : 8$, intervals to compute the integral I from (2.1). Comment on the speed of convergence of Simpson's rule.

5. Let K, T, σ and r be positive constants. Define the function $g : \mathbb{R} \to \mathbb{R}$ as

$$g(x) = \frac{1}{\sqrt{2\pi}} \int_{-\infty}^{b(x)} e^{-\frac{y^2}{2}}\, dy,$$

where

$$b(x) = \left(\ln\left(\frac{x}{K}\right) + \left(r + \frac{\sigma^2}{2}\right) T \right) / \left(\sigma\sqrt{T}\right).$$

Compute $g'(x)$.

Note: This function is related to the Delta of a plain vanilla Call option.

6. Let $h(x)$ be a continuous function such that $\int_{-\infty}^{\infty} |xh(x)| dx$ exists. Define $g(t)$ by

$$g(t) = \int_t^{\infty} (x - t)h(x)\, dx.$$

Show that

$$g''(t) = h(t).$$

Note: The price of a call option can be regarded as a function of the strike price K. By using risk–neutral valuation, we find that

$$\begin{aligned} C(K) &= e^{-rT} E_{RN}[\max(S(T) - K, 0)] \\ &= e^{-rT} \int_{-\infty}^{\infty} \max(x - K, 0) f(x)\, dx \\ &= e^{-rT} \int_{K}^{\infty} (x - K) f(x)\, dx \\ &= \int_{K}^{\infty} (x - K) h(x)\, dx, \end{aligned}$$

where $f(x)$ is the probability density function of $S(T)$ given $S(0)$, and $h(x) = e^{-rT} f(x)$. Then, according to the result of this exercise,

$$\frac{\partial^2 C}{\partial K^2} = e^{-rT} f(K).$$

7. Let I_n^M, I_n^T, and I_n^S be the Midpoint rule, the Trapezoidal rule, and the Simpson's rule approximations to the definite integral $I = \int_a^b f(x)\, dx$ corresponding to a uniform partition of the interval $[a, b]$ into n intervals.

   Show that

$$I_n^S = \frac{2}{3} I_n^M + \frac{1}{3} I_n^T.$$

8. Show that a parallel shift in the zero rate curve generates an identical parallel shift in the instantaneous rate curve, and vice versa.

   In other words, denote by $r_1(0, t)$ and $r_2(0, t)$ two zero rate curves with corresponding instantaneous rate curves $r_1(t)$ and $r_2(t)$. Let $\delta r > 0$ denote an arbitrary positive number. Then,

$$r_2(0, t) = r_1(0, t) + \delta r \quad \text{if and only if} \quad r_2(t) = r_1(t) + \delta r.$$

9. Assume that the continuously compounded zero rate curve is

$$r_c(0, t) = 0.05 + 0.01 \ln\left(1 + \frac{t}{2}\right).$$

   (i) find the instantaneous interest rate curve;

   (ii) compute the corresponding annually compounded zero rate curve;

(iii) compute the corresponding semiannually compounded zero rate curve.

10. The continuously compounded 6-month, 12-month, 18-month, and 24-month zero rates are 5%, 5.25%, 5.35%, and 5.5%, respectively. Find the price of a two year semiannual coupon bond with coupon rate 5%.

11. The continuously compounded 6-month, 12-month, 18-month, and 24-month zero rates are 5%, 5.25%, 5.35%, and 5.5%, respectively. What is the par yield for a 2-year semiannual coupon bond?

12. Assume that the continuously compounded instantaneous interest rate curve has the form

$$r(t) \;=\; 0.05 \;+\; 0.005\ln(1+t), \quad \forall\, t \geq 0.$$

(i) Find the corresponding zero rate curve;

(ii) Compute the 6-month, 12-month, 18-month, and 24-month discount factors;

(iii) Find the price of a two year semiannual coupon bond with coupon rate 5%.

13. Assume that the continuously compounded instantaneous rate curve $r(t)$ is given by

$$r(t) \;=\; \frac{0.05}{1+\exp(-(1+t)^2)}.$$

(i) Use Simpson's Rule to compute the 1–year and 2–year discount factors with six decimal digits accuracy, and compute the 3–year discount factor with eight decimal digits accuracy.

(ii) Find the value of a three year annual coupon bond with coupon rate 5% (and face value 100).

14. The yield of a semiannual coupon bond with 6% coupon rate and 30 months to maturity is 9%. What are the price, duration and convexity of the bond?

15. The yield of a 14 months quarterly coupon bond with 8% coupon rate is 7%. Compute the price, duration, and convexity of the bond.

16. If the coupon rate of a bond goes up, what can be said about the value of the bond and its duration? Give a financial argument.

17. By how much would the price of a ten year zero-coupon bond change if the yield increases by ten basis points? (One percentage point is equal to 100 basis points. Thus, 10 basis points is equal to 0.001.)

18. A five year bond with 3.5 years duration is worth 102. What is the value of the bond if the yield decreases by fifty basis points?

19. Establish the following relationship between duration and convexity:

$$C = D^2 - \frac{\partial D}{\partial y}$$

2.2 Solutions to Chapter 2 Exercises

Problem 1: Compute

$$\lim_{x\to 0} \frac{N(x) - \frac{1}{2} - \frac{x}{\sqrt{2\pi}}}{x^3},$$

where

$$N(x) = \frac{1}{\sqrt{2\pi}} \int_{-\infty}^{x} e^{-\frac{y^2}{2}}\, dy$$

is the cumulative density function of the standard normal variable.

Solution: Recall that

$$N'(x) = \frac{1}{\sqrt{2\pi}} e^{-\frac{x^2}{2}}.$$

Using l'Hôpital's rule, we find that

$$\begin{aligned}
\lim_{x\to 0} \frac{N(x) - \frac{1}{2} - \frac{x}{\sqrt{2\pi}}}{x^3} &= \lim_{x\to 0} \frac{N'(x) - \frac{1}{\sqrt{2\pi}}}{3x^2} \\
&= \lim_{x\to 0} \frac{\frac{1}{\sqrt{2\pi}} e^{-\frac{x^2}{2}} - \frac{1}{\sqrt{2\pi}}}{3x^2} \\
&= \frac{1}{3\sqrt{2\pi}} \lim_{x\to 0} \frac{e^{-\frac{x^2}{2}} - 1}{x^2} \\
&= \frac{1}{3\sqrt{2\pi}} \lim_{x\to 0} \frac{-xe^{-\frac{x^2}{2}}}{2x} \\
&= -\frac{1}{6\sqrt{2\pi}} \lim_{x\to 0} e^{-\frac{x^2}{2}} \\
&= -\frac{1}{6\sqrt{2\pi}}. \quad \square
\end{aligned}$$

Problem 2: Let $f : (0, \infty) \to \mathbb{R}$ denote the Gamma function, i.e., let

$$f(\alpha) = \int_0^\infty x^{\alpha-1}\, e^{-x}\, dx.$$

(i) Show that $f(\alpha)$ is well defined for any $\alpha > 0$, i.e., show that both

$$\int_0^1 x^{\alpha-1}\, e^{-x}\, dx = \lim_{t\searrow 0} \int_t^1 x^{\alpha-1}\, e^{-x}\, dx$$

and
$$\int_1^\infty x^{\alpha-1}\, e^{-x}\, dx \;=\; \lim_{t\to\infty} \int_1^t x^{\alpha-1}\, e^{-x}\, dx$$
exist and are finite.

(ii) Prove, using integration by parts, that $f(\alpha) = (\alpha-1)\, f(\alpha-1)$ for any $\alpha > 1$. Show that $f(1) = 1$ and conclude that, for any $n \geq 1$ positive integer, $f(n) = (n-1)!$.

Solution: (i) Let $\alpha > 0$. Intuitively, note that, as $x \searrow 0$, the function $x^{\alpha-1}e^{-x}$ is on the order of $x^{\alpha-1}$, since $\lim_{x\searrow 0} e^{-x} = 1$. Since
$$\lim_{t\searrow 0} \int_t^1 x^{\alpha-1}\, dx \;=\; \lim_{t\searrow 0} \left.\frac{x^\alpha}{\alpha}\right|_t^1 \;=\; \frac{1}{\alpha} \lim_{t\searrow 0} (1 - t^\alpha) \;=\; \frac{1}{\alpha},$$
it follows that
$$\int_0^1 x^{\alpha-1}\, e^{-x}\, dx \;=\; \lim_{t\searrow 0} \int_t^1 x^{\alpha-1}\, e^{-x}\, dx$$
exists and is finite.

In a similar intuitive way, note that, as $x \to \infty$, the function $x^{\alpha-1}e^{-x}$ is on the order of e^{-x}, since the exponential function dominates any power function at infinity. Since
$$\lim_{t\to\infty} \int_1^t e^{-x}\, dx \;=\; \lim_{t\to\infty} (1 - e^{-t}) \;=\; 1,$$
it follows that
$$\int_1^\infty x^{\alpha-1}\, e^{-x}\, dx \;=\; \lim_{t\to\infty} \int_1^t x^{\alpha-1}\, e^{-x}\, dx \tag{2.2}$$
exists and is finite.

Making these intuitive arguments precise is somewhat more subtle. We include a mathematically rigorous arguments for, e.g., showing that the integral in (2.2) exists and is finite.

By definition, we need to prove that, for any $\epsilon > 0$, there exists $n(\epsilon) > 0$ such that
$$\int_s^\infty x^{\alpha-1}\, e^{-x}\, dx \;<\; \epsilon, \quad \forall\; s > n(\epsilon). \tag{2.3}$$

Note that there exists $N > 0$ such that
$$x^{\alpha-1}\, e^{-x} \;<\; e^{-x/2}, \quad \forall\; x > N, \tag{2.4}$$
since
$$\lim_{x\to\infty} x^{\alpha-1}\, e^{-x/2} \;=\; 0.$$

Also, since $\lim_{x\to\infty} e^{-x/2} = 0$, it follows that, for any $\epsilon > 0$, there exists $m(\epsilon) > 0$ such that

$$2e^{-m(\epsilon)/2} \; < \; \epsilon. \tag{2.5}$$

Choose $n(\epsilon) = \max(m(\epsilon, N))$. From (2.4) and (2.5) we obtain that

$$x^{\alpha-1}\, e^{-x} \; < \; e^{-x/2}, \;\; \forall\, x > n(\epsilon); \tag{2.6}$$

$$2e^{-n(\epsilon)/2} \; < \; \epsilon. \tag{2.7}$$

We can then use (2.6) and (2.7) to show that, for any $s > n(\epsilon)$,

$$\begin{aligned} \int_s^\infty x^{\alpha-1}\, e^{-x}\, dx &= \lim_{t\to\infty} \int_s^t x^{\alpha-1}\, e^{-x}\, dx \\ &< \lim_{t\to\infty} \int_s^t e^{-x/2}\, dx \\ &= \lim_{t\to\infty} \left(-2e^{-t/2} \; + \; 2e^{-s/2}\right) \\ &= 2e^{-s/2} \; < \; 2e^{-n(\epsilon)/2} \\ &< \epsilon, \end{aligned}$$

which is what we wanted to show; cf. (2.3).

(ii) It is easy to see that

$$f(1) \; = \; \int_0^\infty e^{-x}\, dx \; = \; \lim_{t\to\infty} \int_0^t e^{-x}\, dx \; = \; \lim_{t\to\infty} \left(-e^{-t} + 1\right) \; = \; 1.$$

Assume that $\alpha > 1$. By integration by parts, we find that

$$\begin{aligned} f(\alpha) &= \int_0^\infty x^{\alpha-1}\, e^{-x}\, dx \\ &= \lim_{t\to\infty} \int_0^t x^{\alpha-1}\, e^{-x}\, dx \\ &= \lim_{t\to\infty} \left[\left(-x^{\alpha-1}\, e^{-x}\right)\Big|_{x=0}^{x=t} \; + \; (\alpha-1) \int_0^t x^{\alpha-2}\, e^{-x}\, dx\right] \\ &= (\alpha-1) \lim_{t\to\infty} \int_0^t x^{\alpha-2}\, e^{-x}\, dx \\ &= (\alpha-1) f(\alpha-1), \end{aligned}$$

since

$$\begin{aligned} \lim_{x\searrow 0} x^{\alpha-1}\, e^{-x} &= \lim_{x\searrow 0} x^{\alpha-1} \; = \; 0, \quad \text{for } \;\alpha > 1; \\ \lim_{t\to\infty} t^{\alpha-1}\, e^{-t} &= \lim_{t\to\infty} \frac{t^{\alpha-1}}{e^t} \; = \; 0. \end{aligned}$$

For any positive integer $n > 1$, we find that $f(n) = (n-1)f(n-1)$. Since $f(1) = 1$, it follows by induction that

$$f(n) = (n-1)!$$

for any positive integer $n \geq 1$. □

Problem 3: Compute an approximate value of

$$\int_1^3 \sqrt{x}\, e^{-x} dx$$

using the Midpoint rule, the Trapezoidal rule, and Simpson's rule. Start with $n = 4$ intervals, and double the number of intervals until two consecutive approximations are within 10^{-6} of each other.

Solution: The approximate values of the integral found using the Midpoint, Trapezoidal, and Simpson's rules can be found in the table below:

No. Intervals	Midpoint Rule	Trapezoidal Rule	Simpson's Rule
4	0.40715731	0.41075744	0.40835735
8	0.40807542	0.40895737	0.40836940
16	0.40829709	0.40851639	0.40837019
32	0.40835199	0.40840674	0.40837024
64	0.40836569	0.40837937	0.40837024
128	0.40836911	0.40837253	
256	0.40836996	0.40837082	
512	0.40837018	0.40837039	
1024	0.40837023	0.40837028	

The approximate value of the integral is 0.408370, obtained for a 256 intervals partition using the Midpoint rule, for 512 intervals using the Trapezoidal rule, and for 16 intervals using Simpson's rule. □

Problem 4: Let $f : \mathbb{R} \rightarrow \mathbb{R}$ given by

$$f(x) = \frac{x^{5/2}}{1+x^2}.$$

(i) Use Midpoint rule with $tol = 10^{-6}$ to compute an approximation of

$$I = \int_0^1 f(x)\, dx = \int_0^1 \frac{x^{5/2}}{1+x^2}.$$

(ii) Show that $f^{(4)}(x)$ is not bounded on the interval $(0, 1)$.

(iii) Apply Simpson's rule with $n = 2^k$, $k = 2:8$, intervals to compute the integral I. Comment on the speed of convergence of Simpson's rule.

Solution: (i) Using the Midpoint rule, the following approximate values of the integral are obtained:

No. Intervals	Midpoint Rule
4	0.17715737
8	0.17867774
16	0.17904866
32	0.17914062
64	0.17916354
128	0.17916926
256	0.17917070
512	0.17917105

The approximate value of the integral is 0.179171, and is obtained for a partition of the interval $[0, 1]$ using 512 intervals.

(ii) The explanation for this result is that the denominator $1 + x^2$ of $f(x)$ is bounded away from 0, while the fourth derivative of the numerator of $f(x)$ is on the order of $x^{-3/2}$.

By direct computation, it follows that

$$f^{(4)}(x) = x^{-3/2}\frac{g(x)}{(1+x^2)^5},$$

where $g(x)$ is a polynomial with $g(0) \neq 0$. Therefore, $f^{(4)}(x)$ is unbounded as $x \searrow 0$.

(iii) Using Simpson's rule, the following approximate values of the integral are obtained:

No. Intervals	Simpson's Rule
4	0.179155099725056
8	0.179169815603871
16	0.179171055067087
32	0.179171162051226
64	0.179171171372681
128	0.179171172188741
256	0.179171172260393

The approximate value of the integral is 0.17917117, and is obtained for a partition of the interval $[0, 1]$ using 64 intervals. □

Problem 5: Let K, T, σ and r be positive constants. Define the function $g : \mathbb{R} \to \mathbb{R}$ as

$$g(x) = \frac{1}{\sqrt{2\pi}} \int_{-\infty}^{b(x)} e^{-\frac{y^2}{2}} \, dy,$$

where

$$b(x) = \left(\ln\left(\frac{x}{K}\right) + \left(r + \frac{\sigma^2}{2}\right) T \right) / \left(\sigma\sqrt{T}\right).$$

Compute $g'(x)$.

Solution: Recall that

$$\frac{d}{dx}\left(\int_{-\infty}^{b(x)} f(y) \, dy \right) = f(b(x))b'(x),$$

and note that

$$b'(x) = \frac{1}{x\sigma\sqrt{T}}.$$

Therefore,

$$\begin{aligned} g'(x) &= \frac{1}{\sqrt{2\pi}} e^{-\frac{(b(x))^2}{2}} b'(x) \\ &= \frac{1}{x\sigma\sqrt{2\pi T}} \exp\left(-\frac{\left(\ln\left(\frac{x}{K}\right) + \left(r + \frac{\sigma^2}{2}\right) T\right)^2}{2\sigma^2 T} \right). \quad \square \end{aligned}$$

Problem 6: Let $h(x)$ be a continuous function such that $\int_{-\infty}^{\infty} |xh(x)| dx$ exists. Define $g(t)$ by

$$g(t) = \int_t^{\infty} (x - t)h(x) \, dx.$$

Show that

$$g''(t) = h(t). \quad \square$$

Solution: Recall that, if $a(t)$ and $b(t)$ are differentiable functions and if $f(x, t)$ is a continuous function such that $\frac{\partial f}{\partial t}(x, t)$ exists and is continuous, then

$$\frac{d}{dt}\left(\int_{a(t)}^{b(t)} f(x, t) \, dx \right) = \int_{a(t)}^{b(t)} \frac{\partial f}{\partial t}(x, t) \, dx + f(b(t), t)b'(t) - f(a(t), t)a'(t).$$

A similar result can be derived for improper integrals, i.e.,

$$\frac{d}{dt}\left(\int_{a(t)}^{\infty} f(x,t)\,dx \right) \;=\; \int_{a(t)}^{\infty} \frac{\partial f}{\partial t}(x,t)\,dx \;-\; f(a(t),t)a'(t). \qquad (2.8)$$

For our problem,

$$a(t) = t \quad \text{and} \quad f(x,t) = (x-t)h(x), \qquad (2.9)$$

where $h(x)$ is continuous. Then, the function

$$\frac{\partial f}{\partial t}(x,t) \;=\; -\,h(x)$$

is continuous. Note that

$$f(a(t),t) \;=\; f(t,t) \;=\; (t-t)h(t) \;=\; 0. \qquad (2.10)$$

From (2.8–2.10), we conclude that

$$g'(t) \;=\; \frac{d}{dt}\left(\int_{t}^{\infty} (x-t)h(x)\,dx \right) \;=\; -\int_{t}^{\infty} h(x)\,dx.$$

Since

$$\frac{d}{dt}\left(\int_{a(t)}^{\infty} f(x)\,dx \right) \;=\; -\,f(a(t))\,a'(t),$$

it follows that

$$g''(t) \;=\; h(t),$$

which is what we wanted to show. □

Problem 7: Let I_n^M, I_n^T, and I_n^S be the Midpoint rule, the Trapezoidal rule, and the Simpson's rule approximations to the definite integral $I = \int_a^b f(x)\,dx$ corresponding to a uniform partition of the interval $[a,b]$ into n intervals.

Show that

$$I_n^S \;=\; \frac{2}{3}I_n^M \;+\; \frac{1}{3}I_n^T.$$

Solution: Recall that

$$I_n^M \;=\; h\sum_{i=1}^{n} f(x_i);$$

$$I_n^T \;=\; h\left(\frac{f(a_0)}{2} + \frac{f(a_n)}{2}\right) \;+\; h\sum_{i=1}^{n-1} f(a_i);$$

$$I_n^S \;=\; h\left(\frac{f(a_0)}{6} + \frac{f(a_n)}{6}\right) \;+\; \frac{h}{3}\sum_{i=1}^{n-1} f(a_i) \;+\; \frac{2h}{3}\sum_{i=1}^{n} f(x_i), \qquad (2.11)$$

where, $h = \frac{b-a}{n}$, $a_i = a + ih$, $i = 0 : n$, and $x_i = a + \left(i - \frac{1}{2}\right) h$, $i = 1 : n$.
Thus,

$$\begin{aligned} \frac{2}{3} I_n^M + \frac{1}{3} I_n^T &= \frac{2h}{3} \sum_{i=1}^{n} f(x_i) + \frac{h}{3} \left(\frac{f(a_0)}{2} + \frac{f(a_n)}{2} \right) + \frac{h}{3} \sum_{i=1}^{n-1} f(a_i) \quad (2.12) \\ &= I_n^S, \end{aligned}$$

since the right hand sides of (2.11) and (2.12) are the same. □

Problem 8: Show that a parallel shift in the zero rate curve generates an identical parallel shift in the instantaneous rate curve, and vice versa.

In other words, denote by $r_1(0,t)$ and $r_2(0,t)$ two zero rate curves with corresponding instantaneous rate curves $r_1(t)$ and $r_2(t)$. Let $\delta r > 0$ denote an arbitrary positive number. Then,

$$r_2(0,t) = r_1(0,t) + \delta r \quad \text{if and only if} \quad r_2(t) = r_1(t) + \delta r.$$

Solution: (i) Assume that $r_2(t) = r_1(t) + \delta r$. Since

$$r_1(0,t) = \frac{1}{t} \int_0^t r_1(\tau)\, d\tau \quad \text{and} \quad r_2(0,t) = \frac{1}{t} \int_0^t r_2(\tau)\, d\tau,$$

we find that

$$\begin{aligned} r_2(0,t) - r_1(0,t) &= \frac{1}{t} \int_0^t r_2(\tau) - r_1(\tau)\, d\tau \\ &= \frac{1}{t} \int_0^t \delta r\, d\tau \\ &= \frac{1}{t} \cdot (\delta r) t \\ &= \delta r, \end{aligned}$$

which is what we wanted to show.
(ii) Assume that $r_2(0,t) = r_1(0,t) + \delta r$. Since

$$r_1(t) = (t\, r_1(0,t))' \quad \text{and} \quad r_2(t) = (t\, r_2(0,t))',$$

it follows that

$$\begin{aligned} r_2(t) - r_1(t) &= (t r_2(0,t) - t r_1(0,t))' \\ &= (t(r_2(0,t) - r_1(0,t)))' \\ &= (t\, \delta r)' \\ &= \delta r, \end{aligned}$$

which is what we wanted to prove. □

Problem 9: Assume that the continuously compounded zero rate curve is

$$r_c(0,t) \;=\; 0.05 + 0.01 \ln\left(1 + \frac{t}{2}\right).$$

(i) find the instantaneous interest rate curve $r(t)$;
(ii) compute the corresponding annually compounded zero rate curve $r_1(t)$;
(iii) compute the corresponding semiannually compounded zero rate curve $r_2(t)$.

Solution: (i) By definition,

$$r(t) \;=\; (t\, r_c(0,t))',$$

and therefore

$$\begin{aligned} r(t) &= \left(0.05t + 0.01t \ln\left(1 + \frac{t}{2}\right)\right)' \\ &= 0.05 + 0.01 \ln\left(1 + \frac{t}{2}\right) + \frac{0.01t}{2+t}. \end{aligned}$$

(ii) By definition,

$$(1 + r_1(0,t))^t \;=\; \exp(t\, r_c(0,t)),$$

and therefore

$$\begin{aligned} r_1(0,t) &= \exp(r_c(0,t)) - 1 \;=\; \exp\left(0.05 + 0.01 \ln\left(1 + \frac{t}{2}\right)\right) - 1 \\ &= e^{0.05}\left(1 + \frac{t}{2}\right)^{0.01} - 1. \end{aligned}$$

(iii) By definition,

$$\left(1 + \frac{r_2(0,t)}{2}\right)^{2t} \;=\; \exp(t\, r_c(0,t)),$$

and therefore

$$\begin{aligned} r_2(0,t) &= 2\left(\exp\left(\frac{r_c(0,t)}{2}\right) - 1\right) \\ &= 2\left(\exp\left(0.025 + 0.005 \ln\left(1 + \frac{t}{2}\right)\right) - 1\right) \\ &= 2\left(e^{0.025}\left(1 + \frac{t}{2}\right)^{0.005} - 1\right). \quad \Box \end{aligned}$$

Problem 10: The continuously compounded 6-month, 12-month, 18-month, and 24-month zero rates are 5%, 5.25%, 5.35%, and 5.5%, respectively. Find the price of a two year semiannual coupon bond with coupon rate 5%.

Solution: The value B of the semiannual coupon bond is

$$B = \frac{C}{2}\,100\,e^{-r(0,0.5)0.5} + \frac{C}{2}\,100\,e^{-r(0,1)} + \frac{C}{2}\,100\,e^{-r(0,1.5)1.5} + \left(100 + \frac{C}{2}\,100\right) e^{-r(0,2)2},$$

where $C = 0.05$, and $r(0,0.5) = 0.05$, $r(0,1) = 0.0525$, $r(0,1.5) = 0.0535$, $r(0,2) = 0.055$.

The data below refers to the pseudocode from Table 2.5 of [1] for computing the bond price given the zero rate curve.
Input: $n = 4$

$$\text{t\_cash\_flow} = [0.5\ 1\ 1.5\ 2]\,; \quad \text{v\_cash\_flow} = [2.5\ 2.5\ 2.5\ 102.5]\,.$$

The discount factors are

$$disc = [0.97530991 \quad 0.94885432 \quad 0.92288560 \quad 0.89583414],$$

and the price of the bond is $B = 98.940623$. □

Problem 11: The continuously compounded 6-month, 12-month, 18-month, and 24-month zero rates are 5%, 5.25%, 5.35%, and 5.5%, respectively. What is the par yield for a 2-year semiannual coupon bond?

Solution: Par yield is the coupon rate C that makes the value of the bond equal to its face value. For a 2-year semiannual coupon bond, the par yield can be found by solving

$$100 = \frac{C}{2}\,100\,e^{-r(0,0.5)0.5} + \frac{C}{2}\,100\,e^{-r(0,1)} + \frac{C}{2}\,100\,e^{-r(0,1.5)1.5} + \left(100 + \frac{C}{2}\,100\right) e^{-r(0,2)2}.$$

Thus,

$$C = \frac{2(1 - e^{-r(0,2)2})}{e^{-r(0,0.5)0.5} + e^{-r(0,1)} + e^{-r(0,1.5)1.5} + e^{-r(0,2)2}}.$$

For the zero rates given in this problem, the corresponding value of the par yield is $C = 0.05566075$, i.e., 5.566075%. □

Problem 12: Assume that the continuously compounded instantaneous interest rate curve has the form

$$r(t) \;=\; 0.05 \;+\; 0.005\ln(1+t), \quad \forall\, t \geq 0.$$

(i) Find the corresponding zero rate curve;
(ii) Compute the 6-month, 12-month, 18-month, and 24-month discount factors;
(iii) Find the price of a two year semiannual coupon bond with coupon rate 5%.

Solution: (i) Recall that the zero rate curve $r(0,t)$ can be obtained from the instantaneous interest rate curve $r(t)$ as follows:

$$r(0,t) \;=\; \frac{1}{t}\int_0^t r(\tau)\, d\tau, \quad \forall\, t > 0.$$

Then,

$$r(0,t) \;=\; \frac{1}{t}\int_0^t (0.05 \;+\; 0.005\ln(1+\tau))\; d\tau. \qquad (2.13)$$

Using integration by parts, we find that

$$\begin{aligned}
\int \ln(1+\tau)\, d\tau &= (1+\tau)\ln(1+\tau) \;-\; \int (1+\tau)(\ln(1+\tau))'\, d\tau \\
&= (1+\tau)\ln(1+\tau) \;-\; \int (1+\tau)\frac{1}{1+\tau}\, d\tau \\
&= (1+\tau)\ln(1+\tau) \;-\; \int 1\, d\tau \\
&= (1+\tau)\ln(1+\tau) - \tau \;+\; C. \qquad (2.14)
\end{aligned}$$

From (2.13) and (2.14), we conclude that

$$\begin{aligned}
r(0,t) &= \frac{1}{t}\left(\, 0.05t + 0.005((1+t)\ln(1+t) - t)\,\right) \\
&= 0.045 \;+\; 0.005(1+t)\frac{\ln(1+t)}{t}.
\end{aligned}$$

(ii) The 6-month, 12-month, 18-month, and 24-month discount factors are, respectively,

$$\begin{aligned}
\mathrm{disc}(1) &= e^{-r(0,0.5)0.5} \;=\; 0.97478242; \\
\mathrm{disc}(2) &= e^{-r(0,1)} \;=\; 0.94939392; \\
\mathrm{disc}(3) &= e^{-r(0,1.5)1.5} \;=\; 0.92408277; \\
\mathrm{disc}(4) &= e^{-r(0,2)2} \;=\; 0.89899376.
\end{aligned}$$

(iii) The price of the two year semiannual coupon bond with 5% coupon rate is

$$\begin{aligned} B &= \frac{0.05}{2}\,100\; e^{-r(0,0.5)0.5} + \frac{0.05}{2}\,100\; e^{-r(0,1)} + \frac{0.05}{2}\,100\; e^{-r(0,1.5)1.5} \\ &\quad + \left(100 + \frac{0.05}{2}\,100\right) e^{-r(0,2)2} \\ &= 2.5\ \mathrm{disc}(1) + 2.5\ \mathrm{disc}(2) + 2.5\ \mathrm{disc}(3) + 102.5\ \mathrm{disc}(4) \\ &= 99.267508. \quad \square \end{aligned}$$

Problem 13: Assume that the continuously compounded instantaneous rate curve $r(t)$ is given by

$$r(t) = \frac{0.05}{1+\exp(-(1+t)^2)}.$$

(i) Use Simpson's Rule to compute the 1–year and 2–year discount factors with six decimal digits accuracy, and compute the 3–year discount factor with eight decimal digits accuracy.

(ii) Find the value of a three year annual coupon bond with coupon rate 5% (and face value 100).

Solution: (i) Recall that the discount factor corresponding to time t is

$$\mathrm{disc}(t) = \exp\left(-\int_0^t r(\tau)\, d\tau\right).$$

Using Simpson's Rule, we obtain that the 1–year, 2–year, and 3–year discount factors are

$$\mathrm{disc}(1) = 0.956595; \quad \mathrm{disc}(2) = 0.910128; \quad \mathrm{disc}(3) = 0.86574100.$$

(ii) The value of the three year yearly coupon bond is

$$B = 5\ \mathrm{disc}(1) + 5\ \mathrm{disc}(2) + 105\ \mathrm{disc}(3) = 100.236424. \quad \square$$

Problem 14: The yield of a semiannual coupon bond with 6% coupon rate and 30 months to maturity is 9%. What are the price, duration and convexity of the bond?

Solution: The price, duration, and convexity of a bond with cash flows c_i at time t_i, $i = 1:n$, and yield y are

$$B = \sum_{i=1}^{n} c_i e^{-yt_i}; \tag{2.15}$$

$$D = \frac{1}{B}\sum_{i=1}^{n} t_i c_i e^{-yt_i}; \qquad (2.16)$$

$$C = \frac{1}{B}\sum_{i=1}^{n} t_i^2 c_i e^{-yt_i}. \qquad (2.17)$$

The semiannual bond will pay a cash flow of 3 in 6, 12, 18, and 24 months, and will pay 103 at maturity in 30 months.

The data below refers to the pseudocode from Table 2.7 of [1] for computing the price, duration and convexity of a bond given the yield of the bond.

Input: $n = 5$; $y = 0.09$;

$$\text{t\_cash\_flow} = [0.5\ 1\ 1.5\ 2\ 2.5]\,; \quad \text{v\_cash\_flow} = [3\ 3\ 3\ 3\ 103]\,.$$

Output:

Bond price	$B = 92.983915$;
Bond duration	$D = 2.352418$;
Bond convexity	$C = 5.736739$. □

Problem 15: The yield of a 14 months quarterly coupon bond with 8% coupon rate is 7%. Compute the price, duration, and convexity of the bond.

Solution: The quarterly bond will pay a cash flow of 1.75 in 2, 5, 8, and 11 months, and will pay 101.75 at maturity in 14 months. The formulas for the price, duration, and convexity of the bond in terms of the yield y of the bond are similar to those from (2.15–2.17). For example, the price of the bond can be computed as follows:

$$B = 1.75\exp\left(-\frac{2}{12}y\right) + 1.75\exp\left(-\frac{5}{12}y\right) + 1.75\exp\left(-\frac{8}{12}y\right) + 1.75\exp\left(-\frac{11}{12}y\right) + 101.75\exp\left(-\frac{14}{2}y\right).$$

The data below refers to the pseudocode from Table 2.7 of [1] for computing the price, duration and convexity of a bond given the yield of the bond.

Input: $n = 5$; $y = 0.07$;

$$\text{t\_cash\_flow} = \left[\frac{2}{12}\ \frac{5}{12}\ \frac{8}{12}\ \frac{11}{12}\ \frac{14}{12}\right]; \quad \text{v\_cash\_flow} = [2\ 2\ 2\ 2\ 102]\,.$$

Output:

$$\begin{aligned} \text{Bond price} \quad & B = 101.704888; \\ \text{Bond duration} \quad & D = 1.118911; \\ \text{Bond convexity} \quad & C = 1.285705. \quad \square \end{aligned}$$

Problem 16: If the coupon rate of a bond goes up, what can be said about the value of the bond and its duration? Give a financial argument.

Solution: If the coupon rate goes up, the coupon payments increase and therefore the value of the bond increases.

The duration of the bond is the time weighted average of the cash flows, discounted with respect to the yield of the bond. If the coupon rate increases, the duration of the bond decreases. This is due to the fact that the earlier cash flows equal to the coupon payments become a higher fraction of the payment made at maturity, which is equal to the face value of the bond plus a coupon payment, i.e., $\frac{c}{100+c}$ increases as c increases, where c is one coupon payment. $\square$

Problem 17: By how much would the price of a ten year zero-coupon bond change if the yield increases by ten basis points? (One percentage point is equal to 100 basis points. Thus, 10 basis points is equal to 0.001.)

Solution: The duration of a zero–coupon bond is equal to the maturity of the bond, i.e., $D = T = 10$. For a small change

$$\Delta y \;=\; 10\text{bps} \;=\; 0.001$$

in the yield of the bond, the percentage change in the value of a bond can be estimated as follows:

$$\frac{\Delta B}{B} \;\approx\; -\,\Delta y\, D \;=\; -\,0.001 \cdot 10 \;=\; -\,0.01.$$

We conclude that the price of the bond *decreases* by 1%. $\square$

Problem 18: A five year bond with 3.5 years duration is worth 102. What is the value of the bond if the yield decreases by fifty basis points?

Solution: Note that the value of the bond increases, since the yield of the bond decreases.

The return of the bond can be approximated by the duration of the bond multiplied by the parallel shift in the yield curve, with opposite sign, i.e.,

$$\frac{\Delta B}{B} \;\approx\; -\,\Delta y\; D.$$

For $B = 102$, $D = 3.5$ and $\Delta y = -0.005$ (since $1\% = 100$ bps, and therefore $50\text{bps} = 0.5\% = 0.005$), we find that

$$\Delta B \approx -\Delta y\ D\ B = 1.785.$$

The new value of the bond is

$$B_{new} = B + \Delta B = 103.75. \quad \square$$

Problem 19: Establish the following relationship between duration and convexity:

$$C = D^2 - \frac{\partial D}{\partial y}$$

Solution: Recall that

$$D = -\frac{1}{B}\frac{\partial B}{\partial y} \quad \text{and} \quad C = \frac{1}{B}\frac{\partial^2 B}{\partial y^2}.$$

Therefore,

$$\frac{\partial B}{\partial y} = -DB. \tag{2.18}$$

Using Product Rule to differentiate (2.18) with respect to y, we find that

$$\begin{aligned}
\frac{\partial^2 B}{\partial y^2} &= -\frac{\partial D}{\partial y} B - D\frac{\partial B}{\partial y} \\
&= -\frac{\partial D}{\partial y} B - D(-DB) \\
&= -B\frac{\partial D}{\partial y} + BD^2 \\
&= B\left(D^2 - \frac{\partial D}{\partial y}\right).
\end{aligned}$$

We conclude that

$$C = \frac{1}{B}\frac{\partial^2 B}{\partial y^2} = D^2 - \frac{\partial D}{\partial y}. \quad \square$$

Chapter 3

Probability concepts. Black–Scholes formula. Greeks and Hedging.

3.1 Exercises

1. Let k be a positive integer with $2 \leq k \leq 12$. You throw two fair dice. If the sum of the dice is k, you win $w(k)$, or lose 1 otherwise. Find the smallest value of $w(k)$ thats makes the game worth playing.

2. Let X be the number of times you must flip a fair coin before it lands heads. What are $E[X]$ and $\text{var}(X)$?

3. What is the expected number of coin tosses of a fair coin in order to get two heads in a row? What if the coin is biased and the probability of getting heads is p?

4. What is the expected number of tosses in order to get k heads in a row for a biased coin with probability of getting heads equal to p?

5. Over each of three consecutive time intervals of length $\tau = 1/12$, a stock with spot price $S_0 = 40$ at time $t = 0$ will either go up by a factor $u = 1.05$ with probability $p = 0.6$, or down by a factor $d = 0.96$ with probability $1 - p = 0.4$. Compute the expected value and the variance of the stock price at time $T = 3\tau$, i.e., compute $E[S_T]$ and $\text{var}(S_T)$.

6. Calculate the mean and variance of the uniform distribution on the interval [a,b].

7. The density function of the exponential random variable X with parameter $\alpha > 0$ is

$$f(x) = \begin{cases} \alpha\, e^{-\alpha x}, & \text{if } x \geq 0; \\ 0, & \text{if } x < 0. \end{cases}$$

(i) Show that the function $f(x)$ is indeed a density function. It is clear that $f(x) \geq 0$, for any $x \in \mathbb{R}$. Prove that

$$\int_{-\infty}^{\infty} f(x)\, dx = 1.$$

(ii) Show that the expected value and the variance of the exponential random variable X are $E[X] = \frac{1}{\alpha}$ and $\mathrm{var}(X) = \frac{1}{\alpha^2}$.

(iii) Show that the cumulative density of X is

$$F(x) = \begin{cases} 1 - e^{-\alpha x}, & \text{if } \ x \geq 0 \\ 0, & \text{otherwise} \end{cases}$$

(iv) Show that

$$P(X \geq t) = \int_t^{\infty} f(x)\, dx = e^{-\alpha t}.$$

Note: this result is used to show that the exponential variable is memoryless, i.e., $P(X \geq t + s \mid X \geq t) = P(X \geq s)$.

8. Let X be a normally distributed random variable with mean μ and standard deviation $\sigma > 0$. Compute $E[\ |X|\ ]$ and $E[X^2]$.

9. Compute the expected value and variance of the Poisson distribution, i.e., of a random variable X taking only positive integer values with probabilities

$$P(X = k) = \frac{e^{-\lambda}\lambda^k}{k!}, \quad \forall\, k \geq 0,$$

where $\lambda > 0$ is a fixed positive number.

10. Prove that

$$\left(\sum_{i=1}^n x_i\, y_i\right)^2 \leq \left(\sum_{i=1}^n x_i^2\right)\left(\sum_{i=1}^n y_i^2\right), \quad \forall\, x_i, y_i \in \mathbb{R},\ i = 1:n.$$

Note: This inequality is called the Cauchy–Schwarz inequality.

11. Show that

$$\int_a^b f(x)g(x)\,dx \;\le\; \left(\int_a^b f^2(x)\,dx\right)^{\frac{1}{2}} \left(\int_a^b g^2(x)\,dx\right)^{\frac{1}{2}},$$

for any two continuous functions $f, g : \mathbb{R} \to \mathbb{R}$.

Hint: Use the fact that

$$\int_a^b (f(x) + \alpha g(x))^2\,dx \;\ge\; 0, \quad \forall\, \alpha \in \mathbb{R}.$$

12. Use the Black–Scholes formula to price both a put and a call option with strike $K = 45$ expiring in six months on an underlying asset with spot price 50 and volatility 20% paying dividends continuously at 2%, if interest rates are constant at 6%.

13. What is the value of a European Put option with strike $K = 0$? What is the value of a European Call option with strike $K = 0$? How do you hedge a short position in such a call option?

Note: Since no assumptions are made on the evolution of the price of the underlying asset, the Law of One Price and not the Black–Scholes formulas, must be used.

14. Use the formula $\rho(C) = K(T-t)e^{-r(T-t)}N(d_2)$, and the Put–Call parity to show that

$$\rho(P) \;=\; -\,K(T-t)e^{-r(T-t)}N(-d_2).$$

15. Denote by r the risk–free interest rate, assumed to be constant. The value of a European asset–or–nothing call with strike K and maturity T on a lognormally distributed underlying asset with volatility 30% and paying dividends continuously at rate q is

$$C_{AoN} \;=\; Se^{-qT}N(d_1),$$

where

$$d_1 \;=\; \frac{\ln\left(\frac{S}{K}\right) \;+\; \left(r - q + \frac{\sigma^2}{2}\right)T}{\sigma\sqrt{T}}.$$

Find the Delta of the asset–or–nothing call.

16. The sensitivity of the vega of a portfolio with respect to volatility and to the price of the underlying asset are often important to estimate, e.g., for pricing volatility swaps. These two Greeks are called volga and vanna and are defined as follows:

$$\text{volga}(V) = \frac{\partial(\text{vega}(V))}{\partial \sigma} \quad \text{and} \quad \text{vanna}(V) = \frac{\partial(\text{vega}(V))}{\partial S}.$$

It is easy to see that

$$\text{volga}(V) = \frac{\partial^2 V}{\partial \sigma^2} \quad \text{and} \quad \text{vanna}(V) = \frac{\partial^2 V}{\partial S \partial \sigma}.$$

The name volga is the short for “volatility gamma”. Also, vanna can be interpreted as the rate of change of the Delta with respect to the volatility of the underlying asset, i.e.,

$$\text{vanna}(V) = \frac{\partial(\Delta(V))}{\partial \sigma}.$$

(i) Compute the volga and vanna for a plain vanilla European call option on an asset paying dividends continuously at the rate q.

(ii) Use the Put–Call parity to compute the volga and vanna for a plain vanilla European put option.

17. Show that an ATM call on an underlying asset paying dividends continuously at rate q is worth more than an ATM put with the same maturity if and only if $q \leq r$, where r is the constant risk free rate. Use the Put–Call parity, and then use the Black–Scholes formula to prove this result.

18. (i) Show that the Theta of a plain vanilla European call option on a non–dividend–paying asset is always negative.

(ii) Show that the Theta of long dated (i.e., with $T - t$ large) at–the–money calls on an underlying asset paying dividends continuously at a rate equal to the constant risk–free rate, i.e., with $q = r$, may be positive.

19. In the Black–Scholes framework, compute

$$\frac{\partial C}{\partial K} \quad \text{and} \quad \frac{\partial^2 C}{\partial K^2}.$$

Then, use the Put–Call parity to compute

$$\frac{\partial P}{\partial K} \quad \text{and} \quad \frac{\partial^2 P}{\partial K^2}.$$

20. Show that the price of a plain vanilla European call option is a convex function of the strike of the option, i.e., show that

$$\frac{\partial^2 C}{\partial K^2} \geq 0.$$

Hint: Use the "magic of Greeks computations", i.e., the fact that

$$Se^{-q(T-t)}N'(d_1) = Ke^{-r(T-t)}N'(d_2)$$

to obtain that

$$\frac{\partial C}{\partial K} = -e^{-r(T-t)}N(d_2).$$

21. In the Black–Scholes world, for what strike is the value of a straddle minimal? In other words, given S, T, q, r, find the strike K such that $P(K) + C(K)$ is minimal.

22. Compute the Gamma of ATM call options with maturities of fifteen days, three months, and one year, respectively, on a non–dividend–paying underlying asset with spot price 50 and volatility 30%. Assume that interest rates are constant at 5%. What can you infer about the hedging of ATM options with different maturities?

23. (i) The vega of a plain vanilla European call or put is positive, since

$$\text{vega}(C) = \text{vega}(P) = Se^{-q(T-t)}\sqrt{T-t}\,\frac{1}{\sqrt{2\pi}}e^{-\frac{d_1^2}{2}}.$$

Can you give a financial explanation for this?

(ii) Compute the vega of ATM Call options with maturities of fifteen days, three months, and one year, respectively, on a non–dividend–paying underlying asset with spot price 50 and volatility 30%. For simplicity, assume zero interest rates, i.e., $r = 0$.

(iii) If $r = q = 0$, the vega of ATM call and put options is

$$\text{vega}(C) = \text{vega}(P) = S\sqrt{T-t}\,\frac{1}{\sqrt{2\pi}}e^{-\frac{d_1^2}{2}},$$

where $d_1 = \frac{\sigma\sqrt{T-t}}{2}$. Compute the dependence of vega(C) on time to maturity $T - t$, i.e.,

$$\frac{\partial\ (\text{vega}(C))}{\partial(T-t)},$$

and explain the results from part (ii) of the problem.

24. Assume that interest rates are constant and equal to r. Show that, unless the price of a call option C with strike K and maturity T on a non–dividend paying asset with spot price S satisfies the inequality

$$Se^{-qT} - Ke^{-rT} \leq C \leq Se^{-qT},$$

arbitrage opportunities arise.

Show that the value P of the corresponding put option must satisfy the following no–arbitrage condition:

$$Ke^{-rT} - Se^{-qT} \leq P \leq Ke^{-rT}.$$

25. A portfolio containing derivative securities on only one asset has Delta 5000 and Gamma -200. A call on the asset with $\Delta(C) = 0.4$ and $\Gamma(C) = 0.05$, and a put on the same asset, with $\Delta(P) = -0.5$ and $\Gamma(P) = 0.07$ are currently traded. How do you make the portfolio Delta–neutral and Gamma–neutral?

26. You are long 1000 call options with strike 90 and three months to maturity. Assume that the underlying asset has a lognormal distribution with drift $\mu = 0.08$ and volatility $\sigma = 0.2$, and that the spot price of the asset is 92. The risk-free rate is $r = 0.05$. What Delta–hedging position do you need to take?

    Note: For Delta–hedging purposes, it is not necessary to know the drift μ of the underlying asset, since $\Delta(C)$ does not depend on μ.

27. You buy 1000 six months ATM Call options on a non–dividend–paying asset with spot price 100, following a lognormal process with volatility 30%. Assume the interest rates are constant at 5%.

    (i) How much money do you pay for the options?

    (ii) What Delta–hedging position do you have to take?

    (iii) On the next trading day, the asset opens at 98. What is the value of your position (the option and shares position)?

(iv) Had you not Delta–hedged, how much would you have lost due to the decrease in the price of the asset?

28. You hold a portfolio made of a long position in 1000 put options with strike price 25 and maturity of six months, on a non–dividend–paying stock with lognormal distribution with volatility 30%, a long position in 400 shares of the same stock, which has spot price \$20, and \$10,000 in cash. Assume that the risk-free rate is constant at 4%.

    (i) How much is the portfolio worth?

    (ii) How do you adjust the stock position to make the portfolio Delta–neutral?

    (iii) A month later, the spot price of the underlying asset is \$24. What is new value of your portfolio, and how do you adjust the stock position to make the portfolio Delta–neutral?

29. You hold a portfolio with $\Delta(\Pi) = 300$, $\Gamma(\Pi) = 100$, and $\text{vega}(\Pi) = 89$. You can trade in the underlying asset, in a call option with

$$\Delta(C) = 0.2; \quad \Gamma(C) = 0.1; \quad \text{vega}(C) = 0.1,$$

    and in a put option with

$$\Delta(P) = -0.8; \quad \Gamma(P) = 0.3; \quad \text{vega}(P) = 0.2.$$

    What trades do you make to obtain a Δ–, Γ–, and vega–neutral portfolio?

3.2 Solutions to Chapter 3 Exercises

Problem 1: Let k be a positive integer with $2 \leq k \leq 12$. You throw two fair dice. If the sum of the dice is k, you win $w(k)$, or lose 1 otherwise. Find the smallest value of $w(k)$ thats makes the game worth playing.

Solution: Consider the probability space S of all possible outcomes of throwing of the two dice, i.e.,

$$S = \{(x, y) \mid x = 1:6, \ y = 1:6\},$$

where x and y denote the outcomes of the first and second die, respectively. Since the dice are assumed to be fair and the tosses are assumed to be independent of each other, every outcome (x, y) has probability $\frac{1}{36}$ of occurring. Formally, the discrete probability function $P : S \to [0, 1]$ is given by

$$P(x, y) = \frac{1}{36}, \quad \forall \ (x, y) \in S.$$

Let k be a fixed positive integer with $2 \leq k \leq 12$. The value X of your winning (or losses) is the random variable $X : S \to \mathbb{R}$ given by

$$X(x, y) = \begin{cases} w(k), & \text{if } x + y = k; \\ -1, & \text{else.} \end{cases}$$

• If $2 \leq k \leq 7$, then $x + y = k$ if and only if

$$(x, y) \in \{(1, k-1), \ (2, k-2), \ldots, \ (k-1, 1)\}.$$

Note that $x + y = k$ for exactly $k - 1$ of the total of 36 outcomes from S, and $x + y \neq k$ for $36 - (k - 1)$ outcomes from S. Then,

$$\begin{aligned} E[X] &= \sum_{(x,y) \in S} P(x, y) X(x, y) \\ &= \frac{1}{36} \sum_{(x,y) \in S} X(x, y) \\ &= \frac{k-1}{36} w(k) + \frac{36 - (k-1)}{36} (-1) \\ &= \frac{w(k)(k-1) - 37 + k}{36}. \end{aligned} \tag{3.1}$$

The game is worth playing if $E[X] \geq 0$. From (3.1), it follows that, if $2 \leq k \leq 7$, the game should be played if and only if

$$w(k) > \frac{37 - k}{k - 1}.$$

• If $8 \leq k \leq 12$, then $x + y = k$ if and only if

$$(x, y) \in \{(6, k-6),\ (5, k-5), \ldots,\ (k-6, 6)\}.$$

In other words, $x + y = k$ exactly $13 - k$ times. Then,

$$\begin{aligned} E[X] &= \frac{1}{36} \sum_{(x,y)\in S} X(x, y) \\ &= \frac{13-k}{36} w(k) + \frac{36-(13-k)}{36}(-1) \\ &= \frac{w(k)(13-k) - 23 - k}{36}. \end{aligned} \tag{3.2}$$

If $8 \leq k \leq 12$, the game is worth playing if $E[X] \geq 0$, i.e., if and only if

$$w(k) > \frac{23+k}{13-k};$$

cf. (3.2).

The values of $w(k)$ for $k = 2 : 12$ are as follows:

k	2	3	4	5	6	7	8	9	10	11	12
$w(k)$	35	17	11	8	6.2	5	6.2	8	11	17	35

Problem 2: Let X be the number of times you must flip a fair coin until it lands heads. What are $E[X]$ and $\text{var}(X)$?

Solution 1: If the first coin toss is heads (which happens with probability $\frac{1}{2}$), then $X = 1$. If the first coin toss is tails (which also happens with probability $\frac{1}{2}$), then the coin tossing process resets and the number of steps before the coin lands heads will be 1 plus the expected number of coin tosses until the coin lands heads. In other words,

$$E[X] = \frac{1}{2} + \frac{1}{2}(1 + E[X]).$$

We conclude that $E[X] = 2$.

Solution 2: Another way of computing $E[X]$ is as follows: The coin will first land heads in the k-th toss, which corresponds to $X = k$, for a coin toss sequence of $T\ T \ldots T\ H$, i.e., the first $k-1$ tosses are tails, followed by heads once. The probability of this coin toss sequence occuring is $P(T)^{k-1}P(H) = \frac{1}{2^k}$. Then,

$$E[X] = \sum_{k=1}^{\infty} \frac{k}{2^k}. \tag{3.3}$$

Recall that

$$T(n,1,x) = \sum_{k=1}^{n} kx^k = \frac{x-(n+1)x^{n+1}+nx^{n+2}}{(1-x)^2}. \tag{3.4}$$

By letting $x = \frac{1}{2}$ in (3.4), we find that

$$\sum_{k=1}^{n} \frac{k}{2^k} = 2 - \frac{n+1}{2^{n-1}} + \frac{n}{2^n}. \tag{3.5}$$

From (3.3) and (3.5) we conclude that

$$E[X] = \sum_{k=1}^{\infty} \frac{k}{2^k} = \lim_{n\to\infty} \sum_{k=1}^{n} \frac{k}{2^k} = 2.$$

Similarly,

$$E[X^2] = \sum_{k=1}^{\infty} \frac{k^2}{2^k} = \lim_{n\to\infty} \sum_{k=1}^{n} \frac{k^2}{2^k}.$$

Since

$$\begin{aligned} T(n,2,x) &= \sum_{k=1}^{n} k^2\, x^k \\ &= \frac{x+x^2-(n+1)^2x^{n+1}+(2n^2+2n-1)x^{n+2}-n^2x^{n+3}}{(1-x)^3}, \end{aligned}$$

we find that

$$E[X^2] = \lim_{n\to\infty} \sum_{k=1}^{n} \frac{k^2}{2^k} = \lim_{n\to\infty} T\left(n,2,\frac{1}{2}\right) = 6.$$

Therefore,

$$\mathrm{var}(X) = E[X^2] - (E[X])^2 = 2. \quad \square$$

Problem 3: What is the expected number of coin tosses of a fair coin in order to get two heads in a row? What if the coin is biased and the probability of getting heads is p?

Solution: If p is the probability of the coin toss resulting in heads, then the probability of the coin toss resulting in tails is $1-p$.

The outcomes of the first two tosses are as follows:

• If the first toss is tails, which happens with probability $1-p$, then the process resets and the expected number of tosses increases by 1.

• If the first toss is heads, and if the second toss is also heads, which happens with probability p^2, then two consecutive heads were obtained after two tosses.

• If the first toss is heads, and if the second toss is tails, which happens with probability $p(1-p)$, then the process resets and the expected number of tosses increases by 2.

If $E[X]$ denotes the expected number of tosses in order to get two heads in a row, we conclude that

$$E[X] \;=\; (1-p)(1+E[X]) \;+\; 2p^2 + p(1-p)(2+E[X]). \qquad (3.6)$$

We solve (3.6) for $E[X]$ and obtain that

$$E[X] \;=\; \frac{1+p}{p^2}. \qquad (3.7)$$

For an unbiased coin, i.e., for $p=\frac{1}{2}$, we find from (3.7) that $E[X]=6$, and therefore the expected number of coin tosses to obtain two heads in a row is 6. □

Problem 4: What is the expected number of tosses in order to get n heads in a row for a biased coin with probability of getting heads equal to p?

Solution: The probability that the first n throws are all heads is p^n. If the first k throws are heads and the $(k+1)$–th throw is tails, which happens with probability $p^k(1-p)$, then the process resets after the $k+1$ steps; here, $k=0:(n-1)$.

If $x(n)$ denotes the expected number of tosses in order to get n heads in a row, it follows that

$$\begin{aligned} x(n) &= np^n \;+\; \sum_{k=0}^{n-1} p^k(1-p)(k+1+x(n)) \\ &= np^n \;+\; (1-p)\left(\sum_{k=0}^{n-1} p^k + \sum_{k=1}^{n-1} kp^k\right) \;+\; x(n)(1-p)\sum_{k=0}^{n-1} p^k. \end{aligned}$$

Recall that

$$\sum_{k=0}^{n-1} p^k \;=\; \frac{1-p^n}{1-p}; \quad \sum_{k=1}^{n-1} kp^k \;=\; \frac{p-np^n+(n-1)p^{n+1}}{(1-p)^2}.$$

Then, we find that

$$\begin{aligned} x(n) &= np^n + \frac{1-(n+1)p^n + np^{n+1}}{1-p} + x(n)(1-p)\frac{1-p^n}{1-p} \\ &= \frac{1-p^n}{1-p} + x(n)(1-p^n), \end{aligned}$$

and therefore

$$x(n) = \frac{1-p^n}{p^n(1-p)}.$$

We conclude that the expected number of tosses in order to get n heads in a row for a biased coin with probability of getting heads equal to p is

$$\frac{1-p^n}{p^n(1-p)}. \tag{3.8}$$

If the coin were unbiased, we let $p = \frac{1}{2}$ in (3.8) and find out that the expected number of tosses to get n heads in a row is $2^{n+1} - 2$. □

Problem 5: Over each of three consecutive time intervals of length $\tau = 1/12$, the price of a stock with spot price $S_0 = 40$ at time $t = 0$ will either go up by a factor $u = 1.05$ with probability $p = 0.6$, or down by a factor $d = 0.96$ with probability $1 - p = 0.4$. Compute the expected value and the variance of the stock price at time $T = 3\tau$, i.e., compute $E[S_T]$ and $\mathrm{var}(S_T)$.

Solution: The probability space S is the set of all different paths that the stock could follow three consecutive time intervals, i.e.,

$$S = \{UUU, UUD, UDU, UDD, DUU, DUD, DDU, DDD\},$$

where U represents an "up" move and D represents a "down" move.

The value S_T of the stock at time T is a random variable defined on S, and is given by

$$\begin{aligned} S_T(UUU) = S_0u^3; \qquad S_T(DDD) = S_0d^3; \\ S_T(UUD) = S_T(UDU) = S_T(DUU) = S_0u^2d; \\ S_T(UDD) = S_T(DUD) = S_T(DDU) = S_0ud^2. \end{aligned}$$

Note that

$$\begin{aligned} P(UUU) = p^3; \qquad P(DDD) = (1-p)^3; \\ P(UUD) = P(UDU) = P(DUU) = p^2(1-p); \\ P(UDD) = P(DUD) = P(DDU) = p(1-p)^2. \end{aligned}$$

We conclude that

$$\begin{aligned} E[S_T] &= S_0u^3 \cdot p^3 + 3S_0u^2d \cdot p^2(1-p) + 3S_0ud^2 \cdot p(1-p)^2 \\ &\quad + S_0d^3 \cdot (1-p)^3 \\ &= 41.7036; \\ E[(S_T)^2] &= (S_0u^3)^2 \cdot p^3 + 3(S_0u^2d)^2 \cdot p^2(1-p) + 3(S_0ud^2)^2 \cdot p(1-p)^2 \\ &\quad + (S_0d^3)^2 \cdot (1-p)^3 = 1749.0762; \\ \mathrm{var}(S_T) &= E[(S_T)^2] - (E[S_T])^2 = 9.8835. \quad \square \end{aligned}$$

Problem 6 Calculate the mean and variance of the uniform distribution on the interval [a,b].

Solution: The probability density function of the uniform distribution U on the interval $[a, b]$ is the constant function

$$f(x) = \frac{1}{b-a}, \quad \forall\, x \in [a, b].$$

Then,

$$\begin{aligned} E[U] &= \int_a^b x f(x)\, dx = \frac{1}{b-a} \int_a^b x\, dx = \frac{b+a}{2}; \\ \mathrm{var}(U) &= E\left[(U - E[U])^2\right] = E\left[\left(U - \frac{b+a}{2}\right)^2\right] \\ &= \frac{1}{b-a} \int_a^b \left(x - \frac{b+a}{2}\right)^2 dx = \frac{1}{b-a} \cdot \frac{1}{3} \left(x - \frac{b+a}{2}\right)^3 \Bigg|_{x=a}^{x=b} \\ &= \frac{(b-a)^2}{12}. \quad \square \end{aligned}$$

Problem 7: The density function of the exponential random variable X with parameter $\alpha > 0$ is

$$f(x) = \begin{cases} \alpha\, e^{-\alpha x}, & \text{if } x \geq 0; \\ 0, & \text{if } x < 0. \end{cases}$$

(i) Show that the function $f(x)$ is indeed a density function. It is clear that $f(x) \geq 0$, for any $x \in \mathbb{R}$. Prove that

$$\int_{-\infty}^{\infty} f(x)\, dx = 1.$$

(ii) Show that the expected value and the variance of the exponential random variable X are $E[X] = \frac{1}{\alpha}$ and $\mathrm{var}(X) = \frac{1}{\alpha^2}$.

(iii) Show that the cumulative density of X is

$$F(x) = \begin{cases} 1 - e^{-\alpha x}, & \text{if } x \geq 0 \\ 0, & \text{otherwise} \end{cases}$$

(iv) Show that

$$P(X \geq t) = \int_t^\infty f(x)\,dx = e^{-\alpha t}, \quad \forall\, t \geq 0.$$

Note: this result is used to show that the exponential variable is memoryless, i.e., $P(X \geq t + s \mid X \geq t) = P(X \geq s)$.

Solution: (i) It is easy to see that

$$\begin{aligned} \int_{-\infty}^\infty f(x)\,dx &= \int_0^\infty \alpha\, e^{-\alpha x}\,dx = \lim_{t\to\infty} \int_0^t \alpha\, e^{-\alpha x}\,dx \\ &= \lim_{t\to\infty} \left(-e^{-\alpha x}\right)\Big|_{x=0}^{x=t} = \lim_{t\to\infty}(1 - e^{-\alpha t}) = 1. \end{aligned}$$

(ii) By integration by parts we find that

$$\begin{aligned} \int x e^{-\alpha x}\,dx &= -\frac{x e^{-\alpha x}}{\alpha} + \frac{1}{\alpha}\int e^{-\alpha x}\,dx = -\frac{x e^{-\alpha x}}{\alpha} - \frac{e^{-\alpha x}}{\alpha^2}; \\ \int x^2 e^{-\alpha x}\,dx &= -\frac{x^2 e^{-\alpha x}}{\alpha} + \frac{2}{\alpha}\int x e^{-\alpha x}\,dx \\ &= -\frac{x^2 e^{-\alpha x}}{\alpha} - \frac{2x e^{-\alpha x}}{\alpha^2} - \frac{2e^{-\alpha x}}{\alpha^3}. \end{aligned}$$

Then,

$$\begin{aligned} E[X] &= \int_{-\infty}^\infty x f(x)\,dx = \alpha \int_0^\infty x e^{-\alpha x}\,dx = \alpha \lim_{t\to\infty}\int_0^t x e^{-\alpha x}\,dx \\ &= \alpha \lim_{t\to\infty}\left(-\frac{x e^{-\alpha x}}{\alpha} - \frac{e^{-\alpha x}}{\alpha^2}\right)\Big|_0^t = \alpha \lim_{t\to\infty}\left(-\frac{t e^{-\alpha t}}{\alpha} - \frac{e^{-\alpha t}}{\alpha^2} + \frac{1}{\alpha^2}\right) \\ &= \frac{1}{\alpha}; \\ E[X^2] &= \int_{-\infty}^\infty x^2 f(x)\,dx = \alpha \int_0^\infty x^2 e^{-\alpha x}\,dx = \alpha \lim_{t\to\infty}\int_0^t x^2 e^{-\alpha x}\,dx \\ &= \alpha \lim_{t\to\infty}\left(-\frac{x^2 e^{-\alpha x}}{\alpha} - \frac{2x e^{-\alpha x}}{\alpha^2} - \frac{2e^{-\alpha x}}{\alpha^3}\right)\Big|_0^t \end{aligned}$$

$$= \alpha \lim_{t\to\infty} \left(-\frac{t^2 e^{-\alpha t}}{\alpha} - \frac{2te^{-\alpha t}}{\alpha^2} - \frac{2e^{-\alpha t}}{\alpha^3} + \frac{2}{\alpha^3} \right)$$
$$= \frac{2}{\alpha^2}.$$

Therefore,

$$\text{var}(X) = E[X^2] - (E[X])^2 = \frac{2}{\alpha^2} - \left(\frac{1}{\alpha}\right)^2 = \frac{1}{\alpha^2}.$$

(iii) If $x < 0$, then $F(x) = \int_{-\infty}^{x} f(s)\, ds = 0$.
If $x \geq 0$, then

$$F(x) = \int_{-\infty}^{x} f(s)\, ds = \int_0^x \alpha\, e^{-\alpha s}\, ds = \left(-e^{-\alpha s}\right)\Big|_0^x = 1 - e^{-\alpha x}.$$

(iv) If $t \geq 0$, then

$$\begin{aligned} P(X \geq t) &= 1 - P(X < t) = 1 - \int_{-\infty}^{t} f(x)\, dx \\ &= 1 - \int_0^t \alpha\, e^{-\alpha x}\, dx = 1 - \left(-e^{-\alpha x}\right)\Big|_0^t \\ &= e^{-\alpha t}. \end{aligned} \tag{3.9}$$

Recall that the conditional probability of A given B is

$$P(A|B) = \frac{P(A \cap B)}{P(B)}. \tag{3.10}$$

Let $s, t \geq 0$. Then, from (3.9) and (3.10), we find that

$$\begin{aligned} P(X \geq t + s \mid X \geq t) &= \frac{P((X \geq t+s) \cap (X \geq t))}{P(X \geq t)} = \frac{P(X \geq t+s)}{P(X \geq t)} \\ &= \frac{e^{-\alpha(t+s)}}{e^{-\alpha t}} = e^{-\alpha s} = P(X \geq s). \quad \square \end{aligned}$$

Problem 8: Let X be a normally distributed random variable with mean μ and standard deviation $\sigma > 0$. Compute $E[\,|X|\,]$ and $E[X^2]$.

Solution: We compute $E[\,|X|\,]$ in terms of the cumulative distribution

$$N(t) = \frac{1}{\sqrt{2\pi}} \int_{-\infty}^{t} e^{-\frac{x^2}{2}}\, dx$$

of the standard normal variable Z.

Note that $X = \mu + \sigma Z$. Then,

$$\begin{aligned}
E[\,|X|\,] &= \frac{1}{\sqrt{2\pi}} \int_{-\infty}^{\infty} |\mu + \sigma z|\, e^{-\frac{z^2}{2}}\, dz \\
&= \frac{1}{\sqrt{2\pi}} \int_{-\infty}^{-\mu/\sigma} -(\mu + \sigma z)\, e^{-\frac{z^2}{2}} dz \;+\; \frac{1}{\sqrt{2\pi}} \int_{-\mu/\sigma}^{\infty} (\mu + \sigma z)\, e^{-\frac{z^2}{2}} dz \\
&= -\mu \frac{1}{\sqrt{2\pi}} \int_{-\infty}^{-\mu/\sigma} e^{-\frac{z^2}{2}}\, dz \;-\; \frac{\sigma}{\sqrt{2\pi}} \int_{-\infty}^{-\mu/\sigma} z e^{-\frac{z^2}{2}}\, dz \\
&\quad + \mu \frac{1}{\sqrt{2\pi}} \int_{-\mu/\sigma}^{\infty} e^{-\frac{z^2}{2}}\, dz \;+\; \frac{\sigma}{\sqrt{2\pi}} \int_{-\mu/\sigma}^{\infty} z e^{-\frac{z^2}{2}}\, dz.
\end{aligned}$$

It is easy to see that

$$\begin{aligned}
\frac{1}{\sqrt{2\pi}} \int_{-\infty}^{-\mu/\sigma} e^{-\frac{z^2}{2}}\, dz &= N\left(-\frac{\mu}{\sigma}\right) \;=\; 1 - N\left(\frac{\mu}{\sigma}\right); \\
\int_{-\infty}^{-\mu/\sigma} z e^{-\frac{z^2}{2}}\, dz &= \left(-e^{-\frac{z^2}{2}}\right)\Big|_{z=-\infty}^{z=-\mu/\sigma} \;=\; -\exp\left(-\frac{\mu^2}{2\sigma^2}\right); \\
\frac{1}{\sqrt{2\pi}} \int_{-\mu/\sigma}^{\infty} e^{-\frac{z^2}{2}}\, dz &= \frac{1}{\sqrt{2\pi}} \int_{-\infty}^{\mu/\sigma} e^{-\frac{y^2}{2}}\, dy \;=\; N\left(\frac{\mu}{\sigma}\right); \\
\int_{-\mu/\sigma}^{\infty} z e^{-\frac{z^2}{2}}\, dz &= \left(-e^{-\frac{z^2}{2}}\right)\Big|_{z=-\mu/\sigma}^{z=\infty} \;=\; \exp\left(-\frac{\mu^2}{2\sigma^2}\right).
\end{aligned}$$

We conclude that

$$\begin{aligned}
E[\,|X|\,] &= -\,\mu\left(1 - N\left(\frac{\mu}{\sigma}\right)\right) \;+\; \frac{\sigma}{\sqrt{2\pi}} \exp\left(-\frac{\mu^2}{2\sigma^2}\right) \\
&\quad + \mu N\left(\frac{\mu}{\sigma}\right) \;+\; \frac{\sigma}{\sqrt{2\pi}} \exp\left(-\frac{\mu^2}{2\sigma^2}\right) \\
&= \mu\left(2N\left(\frac{\mu}{\sigma}\right) - 1\right) \;+\; \sqrt{\frac{2\sigma}{2\pi}} \exp\left(-\frac{\mu^2}{2\sigma^2}\right).
\end{aligned}$$

One way to compute $E[X^2]$ would be to compute the following integral:

$$E[X^2] \;=\; E[(\mu + \sigma Z)^2] \;=\; \frac{1}{\sqrt{2\pi}} \int_{-\infty}^{\infty} (\mu + \sigma z)^2\, e^{-\frac{z^2}{2}}\, dz.$$

While this would provide the correct result, an easier way is to recall that

$$\mathrm{var}(X) \;=\; E[X^2] - (E[X])^2.$$

Since $E[X] = \mu$ and $\mathrm{var}(X) = \sigma^2$, we conclude that

$$E[X^2] \;=\; \mathrm{var}(X) + (E[X])^2 \;=\; \mu^2 + \sigma^2. \quad \square$$

Problem 9: Compute the expected value and variance of the Poisson distribution, i.e., of a random variable X taking only positive integer values with probabilities

$$P(X = k) \;=\; \frac{e^{-\lambda}\lambda^k}{k!}, \quad \forall\, k \geq 0,$$

where $\lambda > 0$ is a fixed positive number.

Solution: We will show that

$$E[X] = \lambda \quad \text{and} \quad \mathrm{var}(X) = \lambda.$$

By definition,

$$E[X] \;=\; \sum_{k=0}^{\infty} P(X = k) \cdot k \;=\; \sum_{k=1}^{\infty} \frac{e^{-\lambda}\lambda^k}{k!}\, k \;=\; e^{-\lambda}\lambda \sum_{k=1}^{\infty} \frac{\lambda^{k-1}}{(k-1)!}. \quad (3.11)$$

Recall that the Taylor series expansion of e^t is

$$e^t \;=\; \sum_{k=0}^{\infty} \frac{t^k}{k!}.$$

Then, it follows that

$$\sum_{k=1}^{\infty} \frac{\lambda^{k-1}}{(k-1)!} \;=\; e^{\lambda}; \quad (3.12)$$

$$\sum_{k=2}^{\infty} \frac{\lambda^{k-2}}{(k-2)!} \;=\; e^{\lambda}. \quad (3.13)$$

From (3.11) and (3.12), we find that $E[X] \;=\; \lambda$.

Similarly,

$$\begin{aligned}
E[X^2] \;&=\; \sum_{k=0}^{\infty} P(X = k) \cdot k^2 \;=\; \sum_{k=1}^{\infty} \frac{e^{-\lambda}\lambda^k}{k!}\, k^2 \;=\; e^{-\lambda} \sum_{k=1}^{\infty} \frac{k\lambda^k}{(k-1)!} \\
&=\; e^{-\lambda} \sum_{k=1}^{\infty} \frac{(k-1)\lambda^k}{(k-1)!} \;+\; e^{-\lambda} \sum_{k=1}^{\infty} \frac{\lambda^k}{(k-1)!} \\
&=\; e^{-\lambda}\lambda^2 \sum_{k=2}^{\infty} \frac{\lambda^{k-2}}{(k-2)!} \;+\; e^{-\lambda}\lambda \sum_{k=1}^{\infty} \frac{\lambda^{k-1}}{(k-1)!} \\
&=\; \lambda^2 + \lambda,
\end{aligned}$$

where (3.12) and (3.13) were used for the last equality.

We conclude that

$$\operatorname{var}(X) = E[X^2] - (E[X])^2 = \lambda. \quad \square$$

Problem 10: Prove that

$$\left(\sum_{i=1}^n x_i \, y_i \right)^2 \leq \left(\sum_{i=1}^n x_i^2 \right) \left(\sum_{i=1}^n y_i^2 \right), \quad \forall \, x_i, y_i \in \mathbb{R}, \; i = 1 : n. \tag{3.14}$$

Solution: Note that

$$\sum_{i=1}^n (x_i + \alpha y_i)^2 \geq 0, \quad \forall \, \alpha \in \mathbb{R}. \tag{3.15}$$

Also,

$$\sum_{i=1}^n (x_i + \alpha y_i)^2 = \alpha^2 \left(\sum_{i=1}^n y_i^2 \right) + 2\alpha \left(\sum_{i=1}^n x_i \, y_i \right) + \left(\sum_{i=1}^n x_i^2 \right). \tag{3.16}$$

From (3.15) and (3.16), it follows that

$$\alpha^2 \left(\sum_{i=1}^n y_i^2 \right) + 2\alpha \left(\sum_{i=1}^n x_i \, y_i \right) + \left(\sum_{i=1}^n x_i^2 \right) \geq 0, \quad \forall \, \alpha \in \mathbb{R}. \tag{3.17}$$

Note that the left hand side of (3.17) is a quadratic polynomial of α. The inequality (3.17) holds true for any real number α if and only if this polynomial has at most one real double root, i.e., if and only if its discriminant is nonpositive:

$$\left(2 \left(\sum_{i=1}^n x_i \, y_i \right) \right)^2 - 4 \left(\sum_{i=1}^n x_i^2 \right) \left(\sum_{i=1}^n y_i^2 \right) \leq 0,$$

which is equivalent to (3.14). $\square$

Problem 11: Show that

$$\int_a^b f(x) g(x) \, dx \leq \left(\int_a^b f^2(x) \, dx \right)^{\frac{1}{2}} \left(\int_a^b g^2(x) \, dx \right)^{\frac{1}{2}},$$

for any two continuous functions $f, g : \mathbb{R} \to \mathbb{R}$.

Solution: Let $\alpha \in \mathbb{R}$ be an arbitrary real number. Note that

$$\begin{aligned}\int_a^b (f(x) + \alpha g(x))^2 dx &= \int_a^b \left(f^2(x) + 2\alpha f(x)g(x) + \alpha^2 g^2(x)\right) dx \\ &= \alpha^2 \int_a^b g^2(x)dx + 2\alpha \int_a^b f(x)g(x)dx + \int_a^b f^2(x)dx \\ &\geq 0, \quad \forall\, \alpha \in \mathbb{R}.\end{aligned}$$

Recall that a quadratic polynomial $P(x) = Ax^2 + Bx + C$ is nonnegative for all real values of x if and only if $P(x)$ has at most one real double root, which happens if and only if $B^2 - 4AC \leq 0$.

For our problem, it follows that

$$\alpha^2 \int_a^b g^2(x)\, dx \; + \; 2\alpha \int_a^b f(x)g(x)\, dx \; + \; \int_a^b f^2(x)\, dx \; \geq \; 0, \quad \forall\, \alpha \in \mathbb{R}$$

if and only if

$$\left(2 \int_a^b f(x)g(x)\, dx\right)^2 \; - \; 4\left(\int_a^b f^2(x)\, dx\right) \left(\int_a^b g^2(x)\, dx\right) \; \leq \; 0,$$

which is equivalent to

$$\int_a^b f(x)g(x)\, dx \; \leq \; \left(\int_a^b f^2(x)\, dx\right)^{\frac{1}{2}} \left(\int_a^b g^2(x)\, dx\right)^{\frac{1}{2}}. \quad \square$$

Problem 12: Use the Black–Scholes formula to price both a put and a call option with strike 45 expiring in six months on an underlying asset with spot price 50 and volatility 20% paying dividends continuously at 2%, if interest rates are constant at 6%.

Solution: Input for the Black–Scholes formula:

$$S = 50; \; K = 45; \; T - t = 0.5; \; \sigma = 0.2; \; q = 0.02; \; r = 0.06.$$

The Black–Scholes price of the call is $C = 6.508363$ and the price of the put is $P = 0.675920$. $\square$

Problem 13: What is the value of a European Put option with strike $K = 0$? What is the value of a European Call option with strike $K = 0$? How do you hedge a short position in such a call option?

Solution: A put option with strike 0 will never be exercised, since it would mean selling the underlying asset for the price $K = 0$. The value of the put option is 0.

A call with strike 0 will always be exercised, since it gives the right to buy one unit of the underlying asset at zero cost. The value of the call at maturity is $V(T) = S(T)$, and therefore $V(0) = e^{-qT}S(0)$. This can be seen by building a portfolio with a long position on the call option and a short position of e^{-qT} shares, or by using risk–neutral pricing:

$$V(0) \;=\; e^{-rT}E_{RN}[S(T)] \;=\; e^{-rT}\cdot e^{(r-q)T}S(0) \;=\; e^{-qT}S(0).$$

A short position in the call option is hedged (statically) by buying one share of the underlying asset. □

Problem 14: Use formula $\rho(C) = K(T-t)e^{-r(T-t)}N(d_2)$ for $\rho(C)$ and the Put–Call parity to show that

$$\rho(P) \;=\; -\,K(T-t)e^{-r(T-t)}N(-d_2).$$

Solution: Recall that

$$\rho(C) \;=\; \frac{\partial C}{\partial r} \quad \text{and} \quad \rho(P) \;=\; \frac{\partial P}{\partial r}.$$

By differentiating with respect to r the Put–Call parity formula

$$P \;+\; Se^{-q(T-t)} \;-\; C \;=\; Ke^{-r(T-t)},$$

we find that

$$\rho(P) \;-\; \rho(C) \;=\; -\,K(T-t)e^{-r(T-t)}.$$

Therefore,

$$\begin{aligned}
\rho(P) &= \rho(C) \;-\; K(T-t)e^{-r(T-t)} \\
&= K(T-t)e^{-r(T-t)}N(d_2) \;-\; K(T-t)e^{-r(T-t)} \\
&= -K(T-t)e^{-r(T-t)}\,(1 - N(d_2)) \\
&= -K(T-t)e^{-r(T-t)}N(-d_2),
\end{aligned}$$

since $1 - N(d_2) \;=\; N(-d_2)$. □

Problem 15: Denote by r the risk–free interest rate, assumed to be constant. The value of a European asset–or–nothing call with strike K and maturity T on a lognormally distributed underlying asset with volatility 30% and paying dividends continuously at rate q is

$$C_{AoN} \;=\; Se^{-qT}N(d_1),$$

where

$$d_1 = \frac{\ln\left(\frac{S}{K}\right) + \left(r - q + \frac{\sigma^2}{2}\right)T}{\sigma\sqrt{T}}.$$

Find the Delta of the asset–or–nothing call.

Solution: Note that

$$\frac{\partial d_1}{\partial S} = \frac{1}{S\sigma\sqrt{T}}.$$

By direct computation, it follows that

$$\begin{aligned}\Delta(C_{AoN}) &= \frac{\partial C_{AoN}}{\partial S} = e^{-qT}N(d_1) + Se^{-qT}N'(d_1)\frac{\partial d_1}{\partial S} \\ &= e^{-qT}N(d_1) + Se^{-qT}\frac{1}{\sqrt{2\pi}}e^{-\frac{d_1^2}{2}} \cdot \frac{1}{S\sigma\sqrt{T}} \\ &= e^{-qT}N(d_1) + \frac{1}{\sigma\sqrt{2\pi T}}e^{-qT-\frac{d_1^2}{2}}. \quad \square\end{aligned}$$

Problem 16: The sensitivity of the vega of a portfolio with respect to volatility and to the price of the underlying asset are often important to estimate, e.g., for pricing volatility swaps. These two Greeks are called volga and vanna and are defined as follows:

$$\text{volga}(V) = \frac{\partial(\text{vega}(V))}{\partial\sigma} \quad \text{and} \quad \text{vanna}(V) = \frac{\partial(\text{vega}(V))}{\partial S}.$$

It is easy to see that

$$\text{volga}(V) = \frac{\partial^2 V}{\partial\sigma^2} \quad \text{and} \quad \text{vanna}(V) = \frac{\partial^2 V}{\partial S\partial\sigma}.$$

The name volga is the short for "volatility gamma". Also, vanna can be interpreted as the rate of change of the Delta with respect to the volatility of the underlying asset, i.e.,

$$\text{vanna}(V) = \frac{\partial(\Delta(V))}{\partial\sigma}.$$

(i) Compute the volga and vanna for a plain vanilla European call option on an asset paying dividends continuously at the rate q.

(ii) Use the Put–Call parity to compute the volga and vanna for a plain vanilla European put option.

Solution: (i) Recall that

$$\begin{aligned} \text{vega}(C) &= Se^{-q(T-t)}\sqrt{T-t}\,\frac{1}{\sqrt{2\pi}}e^{-\frac{d_1^2}{2}}; \\ \Delta(C) &= e^{-q(T-t)}N(d_1), \end{aligned}$$

where

$$\begin{aligned} d_1 &= \frac{\ln\left(\frac{S}{K}\right) + \left(r-q+\frac{\sigma^2}{2}\right)(T-t)}{\sigma\sqrt{T-t}} \\ &= \frac{\ln\left(\frac{S}{K}\right) + (r-q)(T-t)}{\sigma\sqrt{T-t}} + \frac{\sigma\sqrt{T-t}}{2}. \end{aligned}$$

Then,

$$\begin{aligned} \text{volga}(C) &= \frac{\partial(\text{vega}(C))}{\partial\sigma} = -\,Se^{-q(T-t)}\sqrt{T-t}\frac{1}{\sqrt{2\pi}}\,d_1 e^{-\frac{d_1^2}{2}}\frac{\partial d_1}{\partial\sigma}; \\ \text{vanna}(C) &= \frac{\partial(\Delta(C))}{\partial\sigma} = e^{-q(T-t)}N'(d_1)\,\frac{\partial d_1}{\partial\sigma} = e^{-q(T-t)}\,\frac{1}{\sqrt{2\pi}}e^{-\frac{d_1^2}{2}}\,\frac{\partial d_1}{\partial\sigma}. \end{aligned}$$

Note that

$$\begin{aligned} \frac{\partial d_1}{\partial\sigma} &= -\,\frac{\ln\left(\frac{S}{K}\right) + (r-q)(T-t)}{\sigma^2\sqrt{T-t}} + \frac{\sqrt{T-t}}{2} \\ &= -\,\frac{\ln\left(\frac{S}{K}\right) + \left(r-q-\frac{\sigma^2}{2}\right)(T-t)}{\sigma^2\sqrt{T-t}} \\ &= -\,\frac{d_2}{\sigma}. \end{aligned}$$

We conclude that

$$\text{volga}(C) = Se^{-q(T-t)}\sqrt{T-t}\,\frac{1}{\sqrt{2\pi}}e^{-\frac{d_1^2}{2}}\,\frac{d_1 d_2}{\sigma}; \tag{3.18}$$

$$\text{vanna}(C) = -\,e^{-q(T-t)}\,\frac{1}{\sqrt{2\pi}}e^{-\frac{d_1^2}{2}}\,\frac{d_2}{\sigma}. \tag{3.19}$$

(ii) By differentiating with respect to σ the Put–Call parity formula

$$P + Se^{-q(T-t)} - C = Ke^{-r(T-t)},$$

we find that

$$\text{vega}(P) = \frac{\partial P}{\partial\sigma} = \frac{\partial C}{\partial\sigma} = \text{vega}(C).$$

Therefore,

$$\begin{aligned}
\text{volga}(P) &= \frac{\partial(\text{vega}(P))}{\partial\sigma} = \frac{\partial(\text{vega}(C))}{\partial\sigma} = \text{volga}(C);\\
\text{vanna}(P) &= \frac{\partial(\text{vega}(P))}{\partial S} = \frac{\partial(\text{vega}(C))}{\partial S} = \text{vanna}(C),
\end{aligned}$$

where volga(C) and vanna(C) are given by (3.18) and (3.19), respectively. □

Problem 17: Show that an ATM call on an underlying asset paying dividends continuously at rate q is worth more than an ATM put with the same maturity if and only if $q \leq r$, where r is the constant risk free rate. Use the Put–Call parity, and then use the Black–Scholes formula to prove this result.

Solution 1: For at–the–money options, i.e., with $S = K$, the Put–Call parity can be written as

$$\begin{aligned}
C - P &= Se^{-q(T-t)} - Ke^{-r(T-t)}\\
&= Ke^{-q(T-t)} - Ke^{-r(T-t)}\\
&= Ke^{-r(T-t)}\left(e^{(r-q)(T-t)} - 1\right).
\end{aligned}$$

Therefore, $C \geq P$ if and only if $e^{(r-q)(T-t)} \geq 1$, which is equivalent to $r \geq q$.

Solution 2: Alternatively, the Black–Scholes formulas for at–the–money options can be written as

$$\begin{aligned}
C &= Ke^{-q(T-t)}N(d_1) - Ke^{-r(T-t)}N(d_2);\\
P &= Ke^{-r(T-t)}N(-d_2) - Ke^{-q(T-t)}N(-d_1).
\end{aligned}$$

where

$$d_1 = \left(\frac{r-q}{\sigma} + \frac{\sigma}{2}\right)\sqrt{T-t} \quad \text{and} \quad d_2 = \left(\frac{r-q}{\sigma} - \frac{\sigma}{2}\right)\sqrt{T-t}.$$

Then

$$\begin{aligned}
& & C &\geq P\\
&\Longleftrightarrow & e^{-q(T-t)}N(d_1) - e^{-r(T-t)}N(d_2) &\geq e^{-r(T-t)}N(-d_2) - e^{-q(T-t)}N(-d_1)\\
&\Longleftrightarrow & e^{-q(T-t)}(N(d_1) + N(-d_1)) &\geq e^{-r(T-t)}(N(d_2) + N(-d_2))\\
&\Longleftrightarrow & e^{-q(T-t)} &\geq e^{-r(T-t)}\\
&\Longleftrightarrow & r &\geq q,
\end{aligned}$$

since $N(d_1) + N(-d_1) = N(d_2) + N(-d_2) = 1$. □

Problem 18: (i) Show that the Theta of a plain vanilla European call option on a non–dividend–paying asset is always negative.

(ii) Show that the Theta of long dated (i.e., with $T - t$ large) at–the–money calls on an underlying asset paying dividends continuously at a rate equal to the constant risk–free rate, i.e., with $q = r$, may be positive.

Solution: (i) Recall that

$$\Theta(C) = -\frac{S\sigma e^{-q(T-t)}}{2\sqrt{2\pi(T-t)}} e^{-\frac{d_1^2}{2}} + qSe^{-q(T-t)}N(d_1) - rKe^{-r(T-t)}N(d_2).$$

For a non–dividend–paying asset, i.e., for $q = 0$, we find that

$$\Theta(C) = -\frac{S\sigma}{2\sqrt{2\pi(T-t)}} e^{-\frac{d_1^2}{2}} - rKe^{-r(T-t)}N(d_2) < 0.$$

(ii) If $q = r$, the Theta of an ATM call (i.e., with $S = K$) is

$$\begin{aligned} \Theta(C) &= -\frac{K\sigma e^{-r(T-t)}}{2\sqrt{2\pi(T-t)}} e^{-\frac{d_1^2}{2}} + rKe^{-r(T-t)}N(d_1) - rKe^{-r(T-t)}N(d_2) \\ &= Ke^{-r(T-t)}\left(r(N(d_1) - N(d_2)) - \frac{\sigma}{2\sqrt{2\pi(T-t)}} e^{-\frac{d_1^2}{2}} \right), \end{aligned}$$

where

$$d_1 = \frac{\sigma\sqrt{T-t}}{2} \quad \text{and} \quad d_2 = -\frac{\sigma\sqrt{T-t}}{2}.$$

Note that

$$\lim_{(T-t)\to\infty} d_1 = \infty \quad \text{and} \quad \lim_{(T-t)\to\infty} d_2 = -\infty.$$

Then,

$$\lim_{(T-t)\to\infty} N(d_1) = 1 \quad \text{and} \quad \lim_{(T-t)\to\infty} N(d_2) = 0,$$

and therefore

$$\lim_{(T-t)\to\infty} \left(r(N(d_1) - N(d_2)) - \frac{\sigma}{2\sqrt{2\pi(T-t)}} e^{-\frac{d_1^2}{2}} \right) = r.$$

We conclude that, for $T - t$ large enough,

$$\begin{aligned} \Theta(C) &= Ke^{-r(T-t)}\left(r(N(d_1) - N(d_2)) - \frac{\sigma}{2\sqrt{2\pi(T-t)}} e^{-\frac{d_1^2}{2}} \right) \\ &> 0. \end{aligned}$$

We note that the positive value of $\Theta(C)$ is nonetheless small, since

$$\lim_{(T-t)\to\infty} \Theta(C) = \lim_{(T-t)\to\infty} Ke^{-r(T-t)} \cdot r = 0. \quad \square$$

Problem 19: In the Black–Scholes framework, compute

$$\frac{\partial C}{\partial K} \quad \text{and} \quad \frac{\partial^2 C}{\partial K^2}.$$

Then, use the Put–Call parity to compute

$$\frac{\partial P}{\partial K} \quad \text{and} \quad \frac{\partial^2 P}{\partial K^2}.$$

Solution: Recall the Black–Scholes formula

$$C = Se^{-q(T-t)}N(d_1) - Ke^{-r(T-t)}N(d_2),$$

where

$$d_1 = \frac{\ln\left(\frac{S}{K}\right) + \left(r - q + \frac{\sigma^2}{2}\right)(T-t)}{\sigma\sqrt{T-t}};$$

$$d_2 = d_1 - \sigma\sqrt{T-t} = \frac{\ln\left(\frac{S}{K}\right) + \left(r - q - \frac{\sigma^2}{2}\right)(T-t)}{\sigma\sqrt{T-t}}$$

and the "magic of Greeks computations", i.e., the fact that

$$Se^{-q(T-t)}N'(d_1) = Ke^{-r(T-t)}N'(d_2).$$

Then,

$$\begin{aligned}
\frac{\partial C}{\partial K} &= Se^{-q(T-t)}N'(d_1) \cdot \frac{\partial d_1}{\partial K} - Ke^{-r(T-t)}N'(d_2) \cdot \frac{\partial d_2}{\partial K} - e^{-r(T-t)}N(d_2) \\
&= Se^{-q(T-t)}N'(d_1)\left(\frac{\partial d_1}{\partial K} - \frac{\partial d_2}{\partial K}\right) - e^{-r(T-t)}N(d_2) \\
&= -e^{-r(T-t)}N(d_2),
\end{aligned}$$

since $d_1 - d_2 = \sigma\sqrt{T-t}$ and therefore

$$\frac{\partial d_1}{\partial K} - \frac{\partial d_2}{\partial K} = \frac{\partial(d_1 - d_2)}{\partial K} = 0.$$

We showed that

$$\frac{\partial C}{\partial K} = -e^{-r(T-t)}N(d_2). \tag{3.20}$$

Note that

$$\frac{\partial d_2}{\partial K} = -\frac{1}{K\sigma\sqrt{T-t}}. \tag{3.21}$$

By differentiating (3.20) with respect to K, we obtain that

$$\begin{aligned}\frac{\partial^2 C}{\partial K^2} &= -e^{-r(T-t)} N'(d_2) \cdot \frac{\partial d_2}{\partial K} \\ &= -e^{-r(T-t)} \frac{1}{\sqrt{2\pi}} e^{-\frac{d_2^2}{2}} \cdot \left(-\frac{1}{K\sigma\sqrt{T-t}} \right) \\ &= \frac{1}{K\sigma\sqrt{2\pi(T-t)}} \exp\left(-r(T-t) - \frac{d_2^2}{2} \right). \end{aligned} \tag{3.22}$$

By differentiating with respect to K the Put–Call parity formula

$$P + Se^{-q(T-t)} - C = Ke^{-r(T-t)} \tag{3.23}$$

and using (3.20), we find that

$$\begin{aligned}\frac{\partial P}{\partial K} &= \frac{\partial C}{\partial K} + e^{-r(T-t)} \\ &= -e^{-r(T-t)} N(d_2) + e^{-r(T-t)} \\ &= e^{-r(T-t)} (1 - N(d_2)) \\ &= e^{-r(T-t)} N(-d_2),\end{aligned}$$

since $1 - N(a) = N(-a)$ for any a.

By differentiating twice with respect to K the Put–Call parity (3.23) and using (3.22), we conclude that

$$\frac{\partial^2 P}{\partial K^2} = \frac{\partial^2 C}{\partial K^2} = \frac{1}{K\sigma\sqrt{2\pi(T-t)}} \exp\left(-r(T-t) - \frac{d_2^2}{2} \right). \quad \square$$

Problem 20: Show that the price of a plain vanilla European call option is a convex function of the strike of the option, i.e., show that

$$\frac{\partial^2 C}{\partial K^2} \geq 0.$$

Solution: Recall that

$$Se^{-q(T-t)} N'(d_1) = Ke^{-r(T-t)} N'(d_2).$$

By differentiating the Black–Scholes formula

$$C = Se^{-q(T-t)} N(d_1) - Ke^{-r(T-t)} N(d_2)$$

with respect to K, we obtain that

$$\begin{aligned}\frac{\partial C}{\partial K} &= Se^{-q(T-t)}N'(d_1)\frac{\partial d_1}{\partial K} - Ke^{-r(T-t)}N'(d_2)\frac{\partial d_2}{\partial K} - e^{-r(T-t)}N(d_2)\\ &= Se^{-q(T-t)}N'(d_1)\left(\frac{\partial d_1}{\partial K} - \frac{\partial d_2}{\partial K}\right) - e^{-r(T-t)}N(d_2)\\ &= -e^{-r(T-t)}N(d_2), \qquad (3.24)\end{aligned}$$

since $d_1 = d_2 + \sigma\sqrt{T-t}$ and therefore

$$\frac{\partial d_1}{\partial K} = \frac{\partial d_2}{\partial K}.$$

By differentiating (3.24) with respect to K, we find that

$$\frac{\partial^2 C}{\partial K^2} = -e^{-r(T-t)}\,N'(d_2)\,\frac{\partial d_2}{\partial K} = -e^{-r(T-t)}\,\frac{1}{\sqrt{2\pi}}\,e^{\frac{-d_2^2}{2}}\,\frac{\partial d_2}{\partial K}. \qquad (3.25)$$

Note that

$$\begin{aligned}d_2 &= \frac{\ln\left(\frac{S}{K}\right) + \left(r - q - \frac{\sigma^2}{2}\right)(T-t)}{\sigma\sqrt{T-t}}\\ &= -\frac{\ln(K)}{\sigma\sqrt{T-t}} + \frac{\ln(S) + \left(r - q - \frac{\sigma^2}{2}\right)(T-t)}{\sigma\sqrt{T-t}}.\end{aligned}$$

Then

$$\frac{\partial d_2}{\partial K} = -\frac{1}{\sigma K\sqrt{T-t}}, \qquad (3.26)$$

and, from (3.25) and (3.26), we conclude that

$$\frac{\partial^2 C}{\partial K^2} = \frac{1}{\sigma K\sqrt{2\pi(T-t)}}e^{-r(T-t)}\,e^{\frac{-d_2^2}{2}} \geq 0. \quad \square$$

Problem 21: In the Black–Scholes world, for what strike is the value of a straddle minimal? In other words, given S, T, q, r, find the strike K such that $P(K) + C(K)$ is minimal.

Solution: Recall from the Put–Call parity that

$$P(K) = C(K) + Ke^{-rT} - Se^{-qT},$$

where $P(K)$ and $C(K)$ denote the values of the put option and of the call option with strike K, respectively. Then,

$$P(K) + C(K) = 2C(K) + Ke^{-rT} - Se^{-qT}. \qquad (3.27)$$

Since $\frac{\partial C}{\partial K} = -e^{-r(T-t)}N(d_2)$, see (3.24), we find that

$$\begin{aligned}\frac{\partial}{\partial K}(P(K) + C(K)) &= 2\frac{\partial C}{\partial K} + e^{-rT} = -2e^{-r(T-t)}N(d_2) + e^{-rT} \\ &= e^{-rT}(1 - 2N(d_2)). \end{aligned} \tag{3.28}$$

Then,

$$\frac{\partial}{\partial K}(P(K) + C(K)) = 0 \iff N(d_2) = \frac{1}{2} \iff d_2 = 0 \tag{3.29}$$

Since

$$d_2 = \frac{\ln\left(\frac{S}{K}\right) + \left(r - q - \frac{\sigma^2}{2}\right)T}{\sigma\sqrt{T}},$$

it is easy to see that

$$d_2 = 0 \iff K = S\ \exp\left(\left(r - q - \frac{\sigma^2}{2}\right)T\right). \tag{3.30}$$

Let $K_0 = S\ \exp\left(\left(r - q - \frac{\sigma^2}{2}\right)T\right)$. From (3.29) and (3.30), it follows that the point $K = K_0$ is the only critical point of $P(K) + C(K)$.

If $K < K_0$, then

$$\begin{aligned} d_2 &= \frac{\ln\left(\frac{S}{K}\right) + \left(r - q - \frac{\sigma^2}{2}\right)T}{\sigma\sqrt{T}} \\ &> \frac{\ln\left(\frac{S}{K_0}\right) + \left(r - q - \frac{\sigma^2}{2}\right)T}{\sigma\sqrt{T}} \\ &= 0. \end{aligned}$$

Therefore, $N(d_2) > \frac{1}{2}$ and $\frac{\partial}{\partial K}(P(K) + C(K)) < 0$; cf. (3.28). Thus, if $K < K_0$, then $P(K) + C(K)$ is a decreasing function of K.

If $K > K_0$, then $d_2 < 0$, and therefore $N(d_2) < \frac{1}{2}$ and $\frac{\partial}{\partial K}(P(K) + C(K)) > 0$; cf. (3.28). Thus, if $K > K_0$, then $P(K) + C(K)$ is an increasing function of K.

We conclude that the value $P(K) + C(K)$ of a straddle is minimal for a strike equal to K_0, i.e., for

$$K = S\ \exp\left(\left(r - q - \frac{\sigma^2}{2}\right)T\right). \tag{3.31}$$

To obtain the corresponding minimal value of the straddle, note that, if $K = K_0$, then $d_2 = 0$ and $d_1 = d_2 + \sigma\sqrt{T} = \sigma\sqrt{T}$. Using the Black–Scholes formula, we find that

$$\begin{aligned} C(K_0) &= Se^{-qT}N(d_1) - K_0e^{-rT}N(d_2) \\ &= Se^{-qT}N(\sigma\sqrt{T}) - \frac{1}{2}K_0e^{-rT}. \end{aligned} \tag{3.32}$$

From (3.27) and (3.32), we conclude that the minimal value of a straddle corresponding to the strike $K = K_0$ given by (3.31) is equal to

$$\begin{aligned} P(K_0) + C(K_0) &= 2C(K_0) + K_0e^{-rT} - Se^{-qT} \\ &= 2Se^{-qT}N(\sigma\sqrt{T}) - K_0e^{-rT} + K_0e^{-rT} - Se^{-qT} \\ &= Se^{-qT}\left(2N(\sigma\sqrt{T}) - 1\right). \quad \square \end{aligned}$$

Problem 22: Compute the Gamma of ATM call options with maturities of fifteen days, three months, and one year, respectively, on a non–dividend–paying underlying asset with spot price 50 and volatility 30%. Assume that interest rates are constant at 5%. What can you infer about the hedging of ATM options with different maturities?

Solution: The input in the Black–Scholes formula for the Gamma of the call is $S = K = 50$, $\sigma = 0.3$, $r = 0.05$, $q = 0$. For $T = 1/24$ (assuming a 30 days per month count), $T = 1/4$, and $T = 1$, the following values of the Gamma of the ATM call are obtained:

$$\begin{aligned} \Gamma(15 \text{ days}) &= 0.057664; \\ \Gamma(3 \text{ months}) &= 0.052530; \\ \Gamma(1 \text{ year}) &= 0.025296. \end{aligned}$$

We note that Gamma decreases as the maturity of the options increases. This can be seen by plotting the Delta of a call option as a function of spot price, and noticing that the slope of the Delta around the at–the–money point is steeper for shorter maturities. The cost of Delta–hedging ATM options is higher for short dated options, since small changes in the price of the underlying asset lead to higher changes in the Delta of the option, and therefore may require more often hedge rebalancing. $\square$

Problem 23: (i) The vega of a plain vanilla European call or put is positive, since

$$\text{vega}(C) = \text{vega}(P) = Se^{-q(T-t)}\sqrt{T-t}\,\frac{1}{\sqrt{2\pi}}e^{-\frac{d_1^2}{2}}. \tag{3.33}$$

Can you give a financial explanation for this?

(ii) Compute the vega of ATM Call options with maturities of fifteen days, three months, and one year, respectively, on a non–dividend–paying underlying asset with spot price 50 and volatility 30%. For simplicity, assume zero interest rates, i.e., $r = 0$.

(iii) If $r = q = 0$, the vega of ATM call and put options is

$$\text{vega}(C) \;=\; \text{vega}(P) \;=\; S\sqrt{T-t}\,\frac{1}{\sqrt{2\pi}}e^{-\frac{d_1^2}{2}},$$

where $d_1 = \frac{\sigma\sqrt{T-t}}{2}$. Compute the dependence of vega(C) on time to maturity $T - t$, i.e.,

$$\frac{\partial\,(\text{vega}(C))}{\partial(T-t)},$$

and explain the results from part (ii) of the problem.

Solution: (i) The fact the vega of a plain vanilla European call or put is positive means that, all other things being equal, options on underlying assets with higher volatility are more valuable (or more expensive, depending on whether you have a long or short options position). This could be understood as follows: the higher the volatility of the underlying asset, the higher the risk associated with writing options on the asset. Therefore, the premium charged for selling the option will be higher.

If you have a long position in either put or call options you are essentially "long volatility".

(ii) The input in the Black–Scholes formula for the Gamma of the call is $S = K = 50$, $\sigma = 0.3$, $r = q = 0$. For $T = 1/24$, $T = 1/4$, and $T = 1$, the following values of the vega of the ATM call are obtained:

$$\begin{aligned} \text{vega}(15 \text{ days}) &= 4.069779; \\ \text{vega}(3 \text{ months}) &= 9.945546; \\ \text{vega}(1 \text{ year}) &= 19.723967. \end{aligned}$$

(iii) For clarity, let $\tau = T - t$. For $r = q = 0$, we obtain from (3.33) that

$$\text{vega}(C) \;=\; \frac{S\sqrt{\tau}}{\sqrt{2\pi}}e^{-\frac{d_1^2}{2}} \;=\; \frac{S\sqrt{\tau}}{\sqrt{2\pi}}e^{-\frac{\sigma^2\tau}{8}},$$

since, for an ATM option with $r = q = 0$,

$$d_1 \;=\; \frac{\ln\left(\frac{S}{K}\right) + \left(r - q + \frac{\sigma^2}{2}\right)\tau}{\sigma\sqrt{\tau}} \;=\; \frac{\sigma\sqrt{\tau}}{2}.$$

By direct computation, we find that

$$\begin{aligned}\frac{\partial\,(\text{vega}(C))}{\partial\tau} &= \frac{S}{2\sqrt{2\pi\tau}}e^{-\frac{\sigma^2\tau}{8}} - \frac{\sigma^2 S\sqrt{\tau}}{8\sqrt{2\pi}}e^{-\frac{\sigma^2\tau}{8}}\\ &= \frac{S}{2\sqrt{2\pi\tau}}\left(1-\frac{\sigma^2\tau}{4}\right)e^{-\frac{\sigma^2\tau}{8}}.\end{aligned}$$

For $\sigma = 0.3$ and for time to maturity less than one year, i.e., for $\tau \leq 1$, we find that

$$1 - \frac{\sigma^2\tau}{4} \geq 0.9775,$$

and therefore

$$\frac{\partial\,(\text{vega}(C))}{\partial\tau} \geq 0.$$

We conclude that, for options with moderately large time to maturity, the vega is increasing as time to maturity increases. Therefore we expect that

$$\text{vega(1year)} > \text{vega(3months)} > \text{vega(15days)},$$

which is what we previously obtained by direct computation. □

Problem 24: Assume that interest rates are constant and equal to r. Show that, unless the price C of a call option with strike K and maturity T on a non–dividend paying asset with spot price S satisfies the inequality

$$Se^{-qT} - Ke^{-rT} \leq C \leq Se^{-qT}, \tag{3.34}$$

arbitrage opportunities arise.

Show that the value P of the corresponding put option must satisfy the following no–arbitrage condition:

$$Ke^{-rT} - Se^{-qT} \leq P \leq Ke^{-rT}. \tag{3.35}$$

Solution: One way to prove these bounds on the prices of European options is by using the Put–Call parity, i.e, $P + Se^{-qT} - C = Ke^{-rT}$.

To establish the bounds (3.34) on the price of the call, note that

$$C = Se^{-qT} - Ke^{-rT} + P. \tag{3.36}$$

The payoff of the put at time T is $\max(K - S(T), 0)$ which is less than the strike K. The value P of the put at time 0 cannot be more than Ke^{-rT}, the present value at time 0 of K at time T. Also, the value P of the put option must be greater than 0. Thus,

$$0 \leq P \leq Ke^{-rT}, \tag{3.37}$$

and, from (3.36) and (3.37), we obtain that

$$Se^{-qT} - Ke^{-rT} \leq Se^{-qT} - Ke^{-rT} + P = C \leq Se^{-qT}.$$

To establish the bounds (3.35) on the price of the put, note that

$$P = Ke^{-rT} - Se^{-qT} + C. \tag{3.38}$$

The payoff of the call at time T is $\max(S(T) - K, 0)$ which is less than $S(T)$. The value C of the call at time 0 cannot be more than Se^{-qT}, the present value at time 0 of one unit of the underlying asset at time T, if the dividends paid by the asset at rate q are continuously reinvested in the asset. Also, the value C of the call option must be greater than 0. Thus,

$$0 \leq C \leq Se^{-qT}. \tag{3.39}$$

From (3.38) and (3.39) it follows that

$$Ke^{-rT} - Se^{-qT} \leq P \leq Ke^{-rT}.$$

A more insightful way to prove these bounds is to use arbitrage arguments and the Law of One Price.

Consider a portfolio made of a short position in one call option with strike K and maturity T and a long position in e^{-qT} units of the underlying asset. The value of at time 0 of this portfolio is

$$V(0) = Se^{-qT} - C.$$

If the dividends received on the long asset position are invested continuously in buying more units of the underlying asset, the size of the asset position at time T will be 1 unit of the asset. Thus,

$$V(T) = S(T) - C(T) = S(T) - \max(S(T) - K, 0) \leq K,$$

since, if $S(T) > K$, then $V(T) = S(T) - (S(T) - K) = K$, and, if $S(T) \leq K$, then $V(T) = S(T) \leq K$.

From the Generalized Law of One Price we conclude that

$$V(0) = Se^{-qT} - C \leq Ke^{-rT},$$

and therefore $Se^{-qT} - Ke^{-rT} \leq C$, which is the left inequality from (3.34).

All the other inequalities can be proved similarly:

• To establish that $C \leq Se^{-qT}$, show that the payoff at maturity T of a portfolio made of a long position in e^{-qT} units of the underlying asset at time 0 and a short position in the call option is nonnegative for any possible values of $S(T)$;

• To establish that $Ke^{-rT} - Se^{-qT} \leq P$, show that the payoff at maturity T of a portfolio made of a long position in e^{-qT} units of the underlying asset at time 0 and a long position in the put option is greater than K for any possible values of $S(T)$;

• To establish that $P \leq Ke^{-rT}$, show that the payoff at maturity T of a portfolio made of a short position in the put option and a long cash position of Ke^{-rT} at time 0 is nonnegative for any possible values of $S(T)$. □

Problem 25: A portfolio containing derivative securities on only one asset has Delta 5000 and Gamma −200. A call on the asset with $\Delta(C) = 0.4$ and $\Gamma(C) = 0.05$, and a put on the same asset, with $\Delta(P) = -0.5$ and $\Gamma(P) = 0.07$ are currently traded. How do you make the portfolio Delta–neutral and Gamma–neutral?

Solution: Take positions of size x_1 and x_2, respectively, in the call and put options specified above. The value Π of the new portfolio is $\Pi = V + x_1 C + x_2 P$, where V is the value of the original portfolio. This portfolio will be Delta– and Gamma–neutral, provided that x_1 and x_2 are chosen such that

$$\begin{aligned} \Delta(\Pi) &= \Delta(V) + x_1\Delta(C) + x_2\Delta(P) &= 5000 + 0.4x_1 - 0.5x_2 &= 0; \\ \Gamma(\Pi) &= \Gamma(V) + x_1\Gamma(C) + x_2\Gamma(P) &= -200 + 0.05x_1 + 0.07x_2 &= 0. \end{aligned}$$

The solution of this linear system is

$$x_1 = -\frac{250,000}{53} = -4716.98 \quad \text{and} \quad x_2 = \frac{330,000}{53} = 6226.42.$$

To make the initial portfolio as close to Delta– and Gamma–neutral as possible by only trading in the given call and put options, 4717 calls must be sold and 6226 put options must be bought. The Delta and Gamma of the new portfolio are

$$\begin{aligned} \Delta(\Pi) &= \Delta(V) - 4717\Delta(C) + 6226\Delta(P) &= 0.2; \\ \Gamma(\Pi) &= \Gamma(V) + 4717\Gamma(C) + 6226\Gamma(P) &= -0.03. \end{aligned}$$

To understand how well hedged the portfolio Π is, recall that the initial portfolio had $\Delta(V) = 5000$ and $\Gamma(V) = -200$. □

Problem 26: You are long 1000 call options with strike 90 and three months to maturity. Assume that the underlying asset has a lognormal distribution with drift $\mu = 0.08$ and volatility $\sigma = 0.2$, and that the spot price of the asset is 92. The risk-free rate is $r = 0.05$. What Delta–hedging position do you need to take?

Solution: A long call position is Delta–hedged by a short position in the underlying asset. Delta–hedging the long position in 1000 calls is done by shorting

$$1000\Delta(C) = 1000e^{-qT}N(d_1) = 653.50$$

units of the underlying asset, where

$$d_1 = \frac{\ln\left(\frac{S}{K}\right) + \left(r - q + \frac{\sigma^2}{2}\right)T}{\sigma\sqrt{T}},$$

with $S = 92$, $K = 90$, $T = 1/4$, $\sigma = 0.2$, $r = 0.05$, $q = 0$.

Note that, for Delta–hedging purposes, it is not necessary to know the drift μ of the underlying asset, since $\Delta(C)$ does not depend on μ. □

Problem 27: You buy 1000 six months ATM Call options on a non–dividend–paying asset with spot price 100, following a lognormal process with volatility 30%. Assume the interest rates are constant at 5%.

(i) How much money do you pay for the options?

(ii) What Delta–hedging position do you have to take?

(iii) On the next trading day, the asset opens at 98. What is the value of your position (the option and shares position)?

(iv) Had you not Delta–hedged, how much would you have lost due to the decrease in the price of the asset?

Solution: (i) Using the Black–Scholes formula with input $S_1 = K = 100$, $T = 1/2$, $\sigma = 0.3$, $r = 0.05$, $q = 0$, we find that the value of one call option is $C_1 = 9.634870$. Therefore, \$9,634.87 must be paid for 1000 calls.

(ii) The Delta–hedging position for long 1000 calls is short

$$1000\Delta(C) = 1000e^{-qT}N(d_1) = 588.59$$

units of the underlying. Therefore, 589 units of the underlying must be shorted.

(iii) The new spot price and maturity of the option are $S_2 = 98$ and $T_2 = 125/252$ (there are 252 trading days in one year). The value of the call option is \$8.453134 and the value of the portfolio is

$$1000C_2 - 589S_2 = -49268.87.$$

(iv) If the long call position is not Delta–hedged, the loss incurred due to the decrease in the spot price of the underlying asset is

$$1000(C_2 - C_1) = -\$1181.74.$$

For the Delta–hedged portfolio, the loss incurred is

$$(1000C_2 - 589S_2) - (1000C_1 - 589S_1) \;=\; -\,\$3.74.$$

As expected, this loss is much smaller than the loss incurred if the options positions is not hedged ("naked"). □

Problem 28: You hold a portfolio made of a long position in 1000 put options with strike price 25 and maturity of six months, on a non–dividend–paying stock with lognormal distribution with volatility 30%, a long position in 400 shares of the same stock, which has spot price \$20, and \$10,000 in cash. Assume that the risk-free rate is constant at 4%.

(i) How much is the portfolio worth?

(ii) How do you adjust the stock position to make the portfolio Delta–neutral?

(iii) A month later, the spot price of the underlying asset is \$24. What is new value of your portfolio, and how do you adjust the stock position to make the portfolio Delta–neutral?

Solution: (i) The value of the portfolio is

$$1000P(0) \;+\; 400S(0) \;+\; 10000 \;=\; 22927,$$

where $S(0) = 20$ is the spot price of the underlying asset and the value $P(0) = 4.9273$ of the put option is obtained using the Black–Scholes formula.

(ii) The Delta of the put option position is $-1000N(-d_1) = -803$. (Here and in the rest of the problem, the values of Delta are rounded to the nearest integer.) The Delta of the portfolio is

$$-803 + 400 \;=\; -\,403.$$

To obtain a Delta–neutral portfolio, 403 shares must be purchased for \$8,060. The Delta–neutral portfolio will be made of a long position in 1000 put options a long position in 803 shares of the underlying stock, and \$1,940 in cash.

(iii) A month later, the spot price of the underlying asset is $S\left(\frac{1}{12}\right) = 24$ and the put options have five months left until maturity. The Black–Scholes value of the put option is $P\left(\frac{1}{12}\right) = 2.1818$. The cash position has accrued interest and its current value is $1940\,\exp\left(\frac{0.04}{12}\right) = 1946$. The portfolio is worth

$$1000P\left(\frac{1}{12}\right) \;+\; 803S\left(\frac{1}{12}\right) \;+\; 1946 \;=\; 23400.$$

The new Delta of the portfolio is

$$-1000N(-d_1) \;+\; 803 \;=\; 292.$$

To make the portfolio Delta–neutral, you should sell 292 shares. □

Problem 29: You hold a portfolio with $\Delta(\Pi) = 300$, $\Gamma(\Pi) = 100$ and $\mathrm{vega}(\Pi) = 89$. You can trade in the underlying asset, in a call option with

$$\Delta(C) = 0.2; \quad \Gamma(C) = 0.1; \quad \mathrm{vega}(C) = 0.1,$$

and in a put option with

$$\Delta(P) = -0.8; \quad \Gamma(P) = 0.3; \quad \mathrm{vega}(P) = 0.2.$$

What trades do you make to obtain a Δ–, Γ–, and vega–neutral portfolio?

Solution: You can make the portfolio Γ– and vega– neutral by taking positions in the call and put option, respectively. By trading in the underlying asset, the Γ and vega of the portfolio would not change, and the portfolio can be made Δ–neutral.

Formally, let x_1, x_2, and x_3 be the positions in the underlying asset, the call option, and the put option, respectively. The value of the new portfolio is $\Pi_{new} = \Pi + x_1 S + x_2 C + x_3 P$ and therefore

$$\begin{aligned} \Delta(\Pi_{new}) &= \Delta(\Pi) + x_1 + x_2\Delta(C) + x_3\Delta(P); \\ \Gamma(\Pi_{new}) &= \Gamma(\Pi) + x_2\Gamma(C) + x_3\Gamma(P); \\ \mathrm{vega}(\Pi_{new}) &= \mathrm{vega}(\Pi) + x_2\mathrm{vega}(C) + x_3\mathrm{vega}(P). \end{aligned}$$

Then, $\Delta(\Pi_{new}) = \Gamma(\Pi_{new}) = \mathrm{vega}(\Pi_{new}) = 0$ if and only if

$$\begin{cases} x_1 + 0.2x_2 - 0.8x_3 &= -300; \\ 0.1x_2 + 0.3x_3 &= -100; \\ 0.1x_2 + 0.2x_3 &= -89. \end{cases}$$

The solution (rounded to the nearest integer) is $x_1 = -254$, $x_2 = -670$, $x_3 = -110$. In other words, to make the portfolio Δ–, Γ– and vega– neutral, one must short 254 units of the underlying asset and sell 670 call options and 110 put options. □

Chapter 4

Lognormal random variables. Risk–neutral pricing.

4.1 Exercises

1. Let $X_1 = Z$ and $X_2 = -Z$ be two independent random variables, where Z is the standard normal variable. Show that $X_1 + X_2$ is a normal variable of mean 0 and variance 2, i.e., $X_1 + X_2 = \sqrt{2}Z$.

2. Assume that the normal random variables $X_1, X_2, \ldots, X_n$ of mean μ and variance σ^2 are uncorrelated, i.e, $\mathrm{cov}(X_i, X_j) = 0$, for all $1 \leq i \neq j \leq n$. (This happens, e.g., if $X_1, X_2, \ldots, X_n$ are independent.) If $S_n = \frac{1}{n}\sum_{i=1}^{n} X_i$ is the average of the variables X_i, $i = 1 : n$, show that

$$E[S_n] = \mu \quad \text{and} \quad \mathrm{var}(S_n) = \frac{\sigma^2}{n}.$$

3. Assume that we have a one period binomial model for the evolution of the price of an underlying asset between time t and time $t + \delta t$:

   If $S(t)$ is the price of the asset at time t, then the price $S(t + \delta t)$ of the asset at time $t + \delta t$ will be either $S(t)u$, with (risk–neutral) probability p, or $S(t)d$, with probability $1 - p$. Assume that $u > 1$ and $d < 1$.

   Show that

$$E_{RN}[S(t + \delta t)] = (pu + (1 - p)d)\, S(t); \tag{4.1}$$
$$E_{RN}[S^2(t + \delta t)] = (pu^2 + (1 - p)d^2)\, S^2(t). \tag{4.2}$$

4. If the price $S(t)$ of a non–dividend paying asset has lognormal distribution with drift $\mu = r$ and volatility σ, then,

$$\ln\left(\frac{S(t + \delta t)}{S(t)}\right) = \left(r - \frac{\sigma^2}{2}\right)\delta t + \sigma\sqrt{\delta t}Z.$$

Show that

$$E_{RN}[S(t+\delta t)] = e^{r\delta t} S(t); \tag{4.3}$$
$$E_{RN}[S^2(t+\delta t)] = e^{(2r+\sigma^2)\delta t} S^2(t). \tag{4.4}$$

5. The results of the previous two exercises can be used to calibrate a binomial tree model to a lognormally distributed process. This means finding the up and down factors u and d, and the risk–neutral probability p (of going up) such that the values of $E_{RN}[S(t+\delta t)]$ and $E_{RN}[S^2(t+\delta t)]$ given by (4.1) and (4.2) coincide with the values (4.3) and (4.4) for the lognormal model.

   In other words, we are looking for u, d, and p such that

$$pu + (1-p)d = e^{r\delta t} \tag{4.5}$$
$$pu^2 + (1-p)d^2 = e^{(2r+\sigma^2)\delta t} \tag{4.6}$$

   Since there are two constraints and three unknowns, the solution will not be unique.

   (i) Show that (4.5–4.6) are equivalent to

$$p = \frac{e^{r\delta t} - d}{u - d} \tag{4.7}$$
$$\left(e^{r\delta t} - d\right)\left(u - e^{r\delta t}\right) = e^{2r\delta t}\left(e^{\sigma^2\delta t} - 1\right) \tag{4.8}$$

   (ii) Derive the Cox–Ross–Rubinstein parametrization for a binomial tree, by solving (4.7–4.8) with the additional condition that

$$ud = 1.$$

   Show that the solution can be written as

$$p = \frac{e^{r\delta t} - d}{u - d}; \quad u = A + \sqrt{A^2 - 1}; \quad d = A - \sqrt{A^2 - 1},$$

   where

$$A = \frac{1}{2}\left(e^{-r\delta t} + e^{(r+\sigma^2)\delta t}\right).$$

6. Find a binomial tree parametrization for a risk–neutral probability (of going up or down) equal to $\frac{1}{2}$.

In other words, find the up and down factors u and d such that

$$\begin{aligned} pu + (1-p)d &= e^{r\delta t}; \\ pu^2 + (1-p)d^2 &= e^{(2r+\sigma^2)\delta t}, \end{aligned}$$

if $p = \frac{1}{2}$.

7. Assume that an asset with spot price 50 paying dividends continuously at rate $q = 0.02$ has lognormal distribution with mean $\mu = 0.08$ and volatility $\sigma = 0.3$. Assume that the risk–free rates are constant and equal to $r = 0.05$.

   (i) Find 95% and 99% confidence intervals for the spot price of the asset in 15 days, 1 month, 2 months, 6 months, and 1 year.

   (ii) Find 95% and 99% risk–neutral confidence intervals for the spot price of the asset in 15 days, 1 month, 2 months, 6 months, and 1 year, i.e., assuming that the drift of the asset is equal to the risk–free rate.

8. If you play (American[1]) roulette 100 times, betting \$100 on black each time, what is the probability of winning at least \$1000, and what is the probability of losing at least \$1000?

9. Consider a put option with strike 55 and maturity 4 months on a non–dividend paying asset with spot price 60 which follows a lognormal model with drift $\mu = 0.1$ and volatility $\sigma = 0.3$. Assume that the risk-free rate is constant equal to 0.05.

   (i) Find the probability that the put will expire in the money.

   (ii) Find the risk–neutral probability that the put will expire in the money.

   (iii) Compute $N(-d_2)$.

10. (i) Consider an at-the-money call on a non–dividend paying asset; assume the Black-Scholes framework. Show that the Delta of the option is always greater than 0.5.

    (ii) If the underlying pays dividends at the continuous rate q, when is the Delta of an at-the-money call less than 0.5?

[1]American roulette has 18 red slots, 18 black slots, and two green slots (corresponding to 0 and 00). European roulette, also called French roulette, has only one green slot corresponding to 0.

Note: For most cases, the Delta of an at-the-money call option is close to 0.5.

11. Consider a six months plain vanilla European put option with strike 50 on a lognormally distributed underlying asset paying dividends continuously at 2%. Assume that interest rates are constant at 4%.

    Use risk–neutral valuation to write the value of the put as an integral over a finite interval. Find the value of the put option with six decimal digits accuracy using the Midpoint Rule and using Simpson's Rule. Also, compute the Black–Scholes value P_{BS} of the put and report the approximation errors of the numerical integration approximations at each step.

12. Use risk–neutral pricing to price a supershare, i.e., an option that pays $(\max(S(T) - K, 0))^2$ at the maturity of the option. In other words, compute
$$V(0) \;=\; e^{-rT}\; E_{RN}[(\max(S(T) - K, 0))^2],$$
    where the expected value is computed with respect to the risk–neutral distribution of the price $S(T)$ of the underlying asset at maturity T, which is assumed to follow a lognormal process with drift r and volatility σ. Assume that the underlying asset pays no dividends, i.e., $q = 0$.

13. Use risk–neutral pricing to find the value of an option on a non–dividend–paying asset with lognormal distribution if the payoff of the option at maturity is equal to $\max((S(T))^\alpha - K, 0)$. Here, $\alpha > 0$ is a fixed constant.

14. If the price of an asset follows a normal process, i.e., $dS = \mu dt + \sigma dX$, then
$$S(t_2) \;=\; S(t_1) \;+\; \mu(t_2 - t_1) \;+\; \sigma\sqrt{t_2 - t_1}\; Z, \;\; \forall\; 0 < t_1 < t_2.$$
    Assume that the risk free rate is constant and equal to r.

    (i) Use risk neutrality to find the value of a call option with strike K and maturity T, i.e., compute
$$C(0) \;=\; e^{-rT}\; E_{RN}[\max(S(T) - K, 0)],$$
    where the expected value is computed with respect to $S(T)$ given by
$$S(T) \;=\; S(0) + rT + \sigma\sqrt{T}\; Z.$$

(ii) Use the Put–Call parity to find the value of a put option with strike K and maturity T, if the underlying asset follows a normal process as above.

15. (i) What is the value at time 0 of a derivative security on an underlying asset following a lognormal distribution with volatility σ which, at time T, pays $\frac{1}{S(T)}$, where $S(T)$ is the price of the underlying asset at time T? Assume that the risk free interest rates are constant and equal to r, and that the underlying asset pays dividends continuously at rate q.

    (ii) Find the value at time 0 of a derivative security on the same underlying asset with the following payoff at maturity:

$$V(T) = \max\left(\frac{1}{K} - \frac{1}{S(T)}, \ 0\right).$$

4.2 Solutions to Chapter 4 Exercises

Problem 1: Let $X_1 = Z$ and $X_2 = -Z$ be two independent random variables, where Z is the standard normal variable. Show that $X_1 + X_2$ is a normal variable of mean 0 and variance 2, i.e., $X_1 + X_2 = \sqrt{2}Z$.

Solution: Recall that if X_1 and X_2 are independent normal random variables with mean and variance μ_1 and σ_1^2, and μ_2 and σ_2^2, respectively, then $X_1 + X_2$ is a normal variable with mean $\mu_1 + \mu_2$ and variance $\sigma_1^2 + \sigma_2^2$, and

$$X_1 + X_2 \;=\; \mu_1 + \mu_2 \;+\; \sqrt{\sigma_1^2 + \sigma_2^2}\; Z.$$

For $X_1 = Z$ and $X_2 = -Z$, it follows that $\mu_1 = \mu_2 = 0$ and $\sigma_1 = \sigma_2 = 1$. We conclude that

$$E[X] = \mu_1 + \mu_2 = 0 \quad \text{and} \quad \mathrm{var}(X) = \sigma_1^2 + \sigma_2^2 = 2,$$

and therefore

$$X \;=\; X_1 + X_2 \;=\; \sqrt{2}Z. \quad \square$$

Problem 2: Assume that the normal random variables $X_1, X_2, \ldots, X_n$ of mean μ and variance σ^2 are uncorrelated, i.e, $\mathrm{cov}(X_i, X_j) = 0$, for all $1 \le i \neq j \le n$. (This happens, e.g., if $X_1, X_2, \ldots, X_n$ are independent.) If $S_n = \frac{1}{n}\sum_{i=1}^n X_i$ is the average of the variables X_i, $i = 1:n$, show that

$$E[S_n] \;=\; \mu \quad \text{and} \quad \mathrm{var}(S_n) \;=\; \frac{\sigma^2}{n}.$$

Solution: Recall that, for $c_i \in \mathbb{R}$,

$$E\left[\sum_{i=1}^n c_i X_i\right] \;=\; \sum_{i=1}^n c_i E[X_i];$$

$$\mathrm{var}\left(\sum_{i=1}^n c_i X_i\right) \;=\; \sum_{i=1}^n c_i^2 \mathrm{var}(X_i) \;+\; 2 \sum_{1 \le i < j \le n} c_i c_j \mathrm{cov}(X_i, X_j).$$

Therefore,

$$E[S_n] \;=\; E\left[\frac{1}{n}\sum_{i=1}^n X_i\right] \;=\; \frac{1}{n}\sum_{i=1}^n E[X_i] \;=\; \frac{1}{n}\cdot n\mu \;=\; \mu;$$

$$\mathrm{var}(S_n) \;=\; \mathrm{var}\left(\frac{1}{n}\sum_{i=1}^n X_i\right)$$

$$
\begin{aligned}
&= \frac{1}{n^2}\sum_{i=1}^{n} \operatorname{var}(X_i) + \frac{2}{n^2}\sum_{1\le i<j\le n} \operatorname{cov}(X_i, X_j) \\
&= \frac{1}{n^2}\sum_{i=1}^{n} \operatorname{var}(X_i) \;=\; \frac{1}{n^2}\cdot n\sigma^2 \;=\; \frac{\sigma^2}{n},
\end{aligned}
$$

since $\operatorname{cov}(X_i, X_j) = 0$ for all $1 \le i \neq j \le n$. □

Problem 3: Assume that we have a one period binomial model for the evolution of the price of an underlying asset between time t and time $t+\delta t$: If $S(t)$ is the price of the asset at time t, then the price $S(t+\delta t)$ of the asset at time $t+\delta t$ will be either $S(t)u$, with (risk–neutral) probability p, or $S(t)d$, with probability $1-p$. Assume that $u > 1$ and $d < 1$.

Show that

$$
\begin{aligned}
E_{RN}[S(t+\delta t)] &= (pu \;+\; (1-p)d)\; S(t); && (4.9)\\
E_{RN}[S^2(t+\delta t)] &= (pu^2 \;+\; (1-p)d^2)\; S^2(t). && (4.10)
\end{aligned}
$$

Solution: We can regard $S(t+\delta t)$ as a random variable over the probability space $\{U, D\}$ of the possible moves of the price of the asset from time t to time $t+\delta t$ endowed with the risk–neutral probability function $P : \{U, D\} \to [0, 1]$ with $P(U) = p$ and $P(D) = 1-p$. Then $S(t+\delta t)$ is given by

$$
S(t+\delta t)(U) \;=\; S(t)u; \quad S(t+\delta t)(D) \;=\; S(t)d.
$$

Then, by definition,

$$
\begin{aligned}
E_{RN}[S(t+\delta t)] &= P(U)\cdot S(t+\delta t)(U) \;+\; P(D)\cdot S(t+\delta t)(D)\\
&= (pu + (1-p)d)\; S(t);\\
E_{RN}[S^2(t+\delta t)] &= P(U)\cdot (S(t+\delta t)(U))^2 \;+\; P(D)\cdot (S(t+\delta t)(D))^2\\
&= (pu^2 + (1-p)d^2)\; S^2(t). \;\; \square
\end{aligned}
$$

Problem 4: If the price $S(t)$ of a non–dividend paying asset has lognormal distribution with drift r and volatility σ, show that

$$
\begin{aligned}
E_{RN}[S(t+\delta t)] &= e^{r\delta t} S(t); && (4.11)\\
E_{RN}[S^2(t+\delta t)] &= e^{(2r+\sigma^2)\delta t} S^2(t). && (4.12)
\end{aligned}
$$

Solution 1: If the price $S(t)$ of the non–dividend paying asset has lognormal distribution with drift r and volatility σ, then $\frac{S(t+\delta t)}{S(t)}$ is a lognormal variable given by

$$\ln\left(\frac{S(t+\delta t)}{S(t)}\right) = \left(r - \frac{\sigma^2}{2}\right)\delta t + \sigma\sqrt{\delta t}Z.$$

Recall that, if $\ln(Y) = \mu + \widetilde{\sigma}Z$ is a lognormal random variable with parameters μ and $\widetilde{\sigma}$, the expected value and variance of Y are

$$\begin{aligned} E[Y] &= \exp\left(\mu + \frac{\widetilde{\sigma}^2}{2}\right); \\ \text{var}(Y) &= \exp\left(2\mu + \widetilde{\sigma}^2\right)\left(e^{\widetilde{\sigma}^2} - 1\right). \end{aligned}$$

If $Y = \frac{S(t+\delta t)}{S(t)}$, then $\mu = \left(r - \frac{\sigma^2}{2}\right)\delta t$ and $\widetilde{\sigma} = \sigma\sqrt{\delta t}$ and therefore

$$E\left[\frac{S(t+\delta t)}{S(t)}\right] = \exp\left(\left(r - \frac{\sigma^2}{2}\right)\delta t + \frac{(\sigma\sqrt{\delta t})^2}{2}\right) = e^{r\delta t}; \quad (4.13)$$

$$\begin{aligned} \text{var}\left(\frac{S(t+\delta t)}{S(t)}\right) &= \exp\left(2\left(r - \frac{\sigma^2}{2}\right)\delta t + (\sigma\sqrt{\delta t})^2\right)\cdot\left(e^{(\sigma\sqrt{\delta t})^2} - 1\right) \\ &= e^{2r\delta t}\left(e^{\sigma^2\delta t} - 1\right). \quad (4.14) \end{aligned}$$

Note that,

$$E\left[\frac{S(t+\delta t)}{S(t)}\right] = \frac{1}{S(t)}\cdot E[S(t+\delta t)] \quad (4.15)$$

$$\text{var}\left(\frac{S(t+\delta t)}{S(t)}\right) = \frac{1}{S^2(t)}\cdot \text{var}\left(S(t+\delta t)\right). \quad (4.16)$$

From (4.13) and (4.15), and from (4.14) and (4.16), respectively, we conclude that

$$\begin{aligned} E[S(t+\delta t)] &= e^{r\delta t}S(t); \\ \text{var}\left(S(t+\delta t)\right) &= e^{2r\delta t}\left(e^{\sigma^2\delta t} - 1\right)S^2(t). \quad (4.17) \end{aligned}$$

Note that

$$\begin{aligned} \text{var}\left(S(t+\delta t)\right) &= E[S^2(t+\delta t)] - \left(E[S(t+\delta t)]\right)^2 \\ &= E[S^2(t+\delta t)] - e^{2r\delta t}S^2(t). \quad (4.18) \end{aligned}$$

From (4.17) and (4.18) it follows that

$$E[S^2(t+\delta t)] = \text{var}\left(S(t+\delta t)\right) + e^{2r\delta t}S^2(t) = e^{2r\delta t+\sigma^2\delta t}S^2(t),$$

which is what we wanted to show.

Solution 2: Note that $S(t+\delta t)$ can be written as a function of the standard normal variable Z as follows:

$$S(t+\delta t) \;=\; S(t)\exp\left(\left(r-\frac{\sigma^2}{2}\right)\delta t \;+\; \sigma\sqrt{\delta t}Z\right).$$

Then,

$$\begin{aligned}
E[S(t+\delta t)] &= \frac{1}{\sqrt{2\pi}}\int_{-\infty}^{\infty} S(t)\exp\left(\left(r-\frac{\sigma^2}{2}\right)\delta t+\sigma\sqrt{\delta t}\;x\right)\,e^{-\frac{x^2}{2}}\,dx \\
&= S(t)\,\frac{1}{\sqrt{2\pi}}\int_{-\infty}^{\infty}\exp\left(r\delta t-\frac{\sigma^2\delta t}{2}+\sigma\sqrt{\delta t}\;x-\frac{x^2}{2}\right)\,dx \\
&= S(t)e^{r\delta t}\,\frac{1}{\sqrt{2\pi}}\int_{-\infty}^{\infty}\exp\left(-\frac{(x-\sigma\sqrt{\delta t})^2}{2}\right)\,dx \\
&= S(t)e^{r\delta t}\,\frac{1}{\sqrt{2\pi}}\int_{-\infty}^{\infty}e^{-\frac{y^2}{2}}\,dy \\
&= S(t)e^{r\delta t},
\end{aligned}$$

where we used the substitution $y = x-\sigma\sqrt{\delta t}$ and the fact that

$$\frac{1}{\sqrt{2\pi}}\int_{-\infty}^{\infty}e^{-\frac{y^2}{2}}\,dy \;=\; 1,$$

since $\frac{1}{\sqrt{2\pi}}e^{-\frac{y^2}{2}}$ is the density function of the standard normal variable.

Similarly, we obtain that

$$\begin{aligned}
E[S^2(t+\delta t)] &= \frac{1}{\sqrt{2\pi}}\int_{-\infty}^{\infty} S^2(t)\exp\left(2\left(r-\frac{\sigma^2}{2}\right)\delta t+2\sigma\sqrt{\delta t}\;x\right)\,e^{-\frac{x^2}{2}}\,dx \\
&= \frac{1}{\sqrt{2\pi}}\int_{-\infty}^{\infty} S^2(t)\exp\left(2r\delta t-\sigma^2\delta t+2\sigma\sqrt{\delta t}\;x-\frac{x^2}{2}\right)\,dx \\
&= S^2(t)\,\frac{1}{\sqrt{2\pi}}\int_{-\infty}^{\infty}\exp\left(2r\delta t+\sigma^2\delta t\;-\;\frac{(x-2\sigma\sqrt{\delta t})^2}{2}\right)\,dx \\
&= S^2(t)e^{(2r+\sigma^2)\delta t}\,\frac{1}{\sqrt{2\pi}}\int_{-\infty}^{\infty}\exp\left(-\frac{(x-2\sigma\sqrt{\delta t})^2}{2}\right)\,dx \\
&= S^2(t)e^{(2r+\sigma^2)\delta t}\,\frac{1}{\sqrt{2\pi}}\int_{-\infty}^{\infty}e^{-\frac{s^2}{2}}\,ds \\
&= S^2(t)e^{(2r+\sigma^2)\delta t};
\end{aligned}$$

the substitution $s = x-2\sigma\sqrt{\delta t}$ was used above. □

Problem 5: The results of the previous two exercises can be used to calibrate a binomial tree model to a lognormally distributed process. This means finding the up and down factors u and d, and the risk–neutral probability p (of going up) such that the values of $E_{RN}[S(t+\delta t)]$ and $E_{RN}[S^2(t+\delta t)]$ given by (4.9) and (4.10) coincide with the values (4.11) and (4.12) for the lognormal model.

In other words, we are looking for u, d, and p such that

$$pu + (1-p)d = e^{r\delta t}; \tag{4.19}$$

$$pu^2 + (1-p)d^2 = e^{(2r+\sigma^2)\delta t}. \tag{4.20}$$

Since there are two constraints and three unknowns, the solution will not be unique.

(i) Show that (4.19–4.20) are equivalent to

$$p = \frac{e^{r\delta t} - d}{u - d}; \tag{4.21}$$

$$\left(e^{r\delta t} - d\right)\left(u - e^{r\delta t}\right) = e^{2r\delta t}\left(e^{\sigma^2\delta t} - 1\right). \tag{4.22}$$

(ii) Derive the Cox–Ross–Rubinstein parametrization for a binomial tree, by solving (4.21–4.22) with the additional condition that

$$ud = 1.$$

Show that the solution can be written as

$$p = \frac{e^{r\delta t} - d}{u - d}; \quad u = A + \sqrt{A^2 - 1}; \quad d = A - \sqrt{A^2 - 1},$$

where

$$A = \frac{1}{2}\left(e^{-r\delta t} + e^{(r+\sigma^2)\delta t}\right). \tag{4.23}$$

Solution: (i) Formula (4.21) can be obtained by solving the linear equation (4.19) for p.

To obtain (4.22), we first square formula (4.19) to obtain

$$p^2u^2 + 2p(1-p)ud + (1-p)^2d^2 = e^{2r\delta t}$$

and subtract this from (4.20). We find that

$$p(1-p)u^2 - 2p(1-p)ud + p(1-p)d^2 = e^{(2r+\sigma^2)\delta t} - e^{2r\delta t},$$

which can be written as

$$p(1-p)(u-d)^2 = e^{2r\delta t}\left(e^{\sigma^2\delta t} - 1\right). \tag{4.24}$$

Using formula (4.21) for p, it is easy to see that

$$p(1-p) \;=\; \frac{\left(e^{r\delta t}-d\right)\left(u-e^{r\delta t}\right)}{(u-d)^2}. \qquad (4.25)$$

From (4.24) and (4.25), we conclude that

$$\left(e^{r\delta t}-d\right)\left(u-e^{r\delta t}\right) \;=\; e^{2r\delta t}\left(e^{\sigma^2\delta t}-1\right).$$

(ii) By multiplying out (4.22) and using the fact that $ud=1$, we obtain that

$$ue^{r\delta t}-1-e^{2r\delta t}+de^{r\delta t} \;=\; e^{(2r+\sigma^2)\delta t}-e^{2r\delta t}. \qquad (4.26)$$

After canceling out the term $-e^{2r\delta t}$, we divide (4.26) by $e^{r\delta t}$ and obtain

$$u+d-e^{-r\delta t} \;=\; e^{(r+\sigma^2)\delta t},$$

which can be written as

$$u+\frac{1}{u}-\left(e^{-r\delta t}+e^{(r+\sigma^2)\delta t}\right) \;=\; u+\frac{1}{u}-2A \;=\; 0;$$

cf. (4.23) for the definition of A.

In other words, u is a solution of the quadratic equation

$$u^2-2Au+1=0, \qquad (4.27)$$

which has two solutions, $A+\sqrt{A^2-1}$ and $A-\sqrt{A^2-1}$. Since $u>1$, we conclude that

$$u=A+\sqrt{A^2-1};$$

the other solution of the quadratic equation (4.27) corresponds to the value of d, since

$$d \;=\; \frac{1}{u} \;=\; \frac{1}{A+\sqrt{A^2-1}} \;=\; A-\sqrt{A^2-1}. \quad \square$$

Problem 6: Find a binomial tree parametrization for a risk–neutral probability (of going up) equal to $\frac{1}{2}$.

In other words, find the up and down factors u and d such that

$$\begin{aligned} pu\;+\;(1-p)d &= e^{r\delta t}; \\ pu^2\;+\;(1-p)d^2 &= e^{(2r+\sigma^2)\delta t}, \end{aligned}$$

if $p=\frac{1}{2}$.

Solution: It is easy to see that, if $p = \frac{1}{2}$, then

$$\begin{aligned} u + d &= 2e^{r\delta t}; \\ u^2 + d^2 &= 2e^{(2r+\sigma^2)\delta t}, \end{aligned}$$

and therefore

$$ud = \frac{(u+d)^2 - (u^2+d^2)}{2} = 2e^{2r\delta t} - e^{(2r+\sigma^2)\delta t}.$$

Note that u and d are the solutions of $z^2 - (u+d)z + ud = 0$, which is the same as

$$z^2 - 2e^{r\delta t}z + 2e^{2r\delta t} - e^{(2r+\sigma^2)\delta t} = 0. \tag{4.28}$$

We solve (4.28) and find that

$$\begin{aligned} u &= e^{r\delta t}\left(1 + \sqrt{e^{\sigma^2\delta t} - 1}\right); \\ d &= e^{r\delta t}\left(1 - \sqrt{e^{\sigma^2\delta t} - 1}\right), \end{aligned}$$

since $d < u$. □

Problem 7: Assume that an asset with spot price 50 paying dividends continuously at rate $q = 0.02$ has lognormal distribution with mean $\mu = 0.08$ and volatility $\sigma = 0.3$. Assume that the risk–free rates are constant and equal to $r = 0.05$.

(i) Find 95% and 99% confidence intervals for the spot price of the asset in 15 days, 1 month, 2 months, 6 months, and 1 year.

(ii) Find 95% and 99% risk–neutral confidence intervals for the spot price of the asset in 15 days, 1 month, 2 months, 6 months, and 1 year, i.e., assuming that the drift of the asset is equal to the risk–free rate.

Solution: If the asset has lognormal distribution, then

$$\begin{aligned} S(t) &= S(0)\exp\left(\left(\mu - q - \frac{\sigma^2}{2}\right)t + \sigma\sqrt{t}Z\right) \\ &= 50\exp(0.015t + 0.3\sqrt{t}Z). \end{aligned} \tag{4.29}$$

(i) Recall that the 95% and 99% confidence intervals for the standard normal distribution Z are $[-1.95996, 1.95996]$ and $[-2.57583, 2.57583]$, i.e.,

$$P(-1.95996 < Z < 1.95996) = 0.95; \quad P(-2.57583 < Z < 2.57583) = 0.99.$$

The 95% confidence interval for $S(t)$ is

$$\left[50\exp(0.015t - 0.3\sqrt{t}\cdot 1.95996),\ 50\exp(0.015t + 0.3\sqrt{t}\cdot 1.95996)\right].$$

The 99% confidence interval for $S(t)$ is

$$[50\exp(0.015t - 0.3\sqrt{t}\cdot 2.57583),\ 50\exp(0.015t + 0.3\sqrt{t}\cdot 2.57583)].$$

(ii) The *risk–neutral* confidence intervals for the spot price of the asset are found by substituting the risk–free rate r for μ in (4.29) to obtain

$$S_{RN}(t) = 50\exp(-0.015t + 0.3\sqrt{t}Z).$$

The 95% confidence interval for $S_{RN}(t)$ is

$$[50\exp(-0.015t - 0.3\sqrt{t}\cdot 1.95996),\ 50\exp(-0.015t + 0.3\sqrt{t}\cdot 1.95996)].$$

The 99% confidence interval for $S_{RN}(t)$ is

$$[50\exp(-0.015t - 0.3\sqrt{t}\cdot 2.57583),\ 50\exp(-0.015t + 0.3\sqrt{t}\cdot 2.57583)].$$

For $t \in \left\{\frac{1}{24}, \frac{1}{12}, \frac{1}{6}, \frac{1}{2}, 1\right\}$, we obtain the following confidence intervals:

t	95% CI $S(t)$	99% CI $S(t)$	95% CI $S_{RN}(t)$	99% CI $S_{RN}(t)$
15 days	[43.36, 57.76]	[41.45, 60.43]	[43.28, 57.66]	[41.36, 60.30]
1 month	[42.25, 59.32]	[40.05, 62.57]	[42.14, 59.18]	[39.96, 64.41]
2 months	[39.43, 63.72]	[36.56, 68.70]	[39.23, 63.41]	[36.36, 68.37]
6 months	[32.25, 75.35]	[29.17, 86.99]	[32.74, 75.21]	[28.73, 85.70]
1 year	[28.20, 91.38]	[23.44,109.90]	[27.36, 88.68]	[22.75,106.65]

Problem 8: If you play (American) roulette 100 times, betting \$100 on black each time, what is the probability of winning at least \$1000, and what is the probability of losing at least \$1000?

Solution: Recall that an American roulette has 18 red slots, 18 black slots, and two green slots. Therefore, every time you bet \$100 on black, you win \$100 with probability $\frac{9}{19}$ and lose \$100 with probability $\frac{10}{19}$. In other words, if W_i is the value of the winnings in the i–th round of playing, then

$$W_i = \begin{cases} -100, & \text{with probability } \frac{10}{19}; \\ 100, & \text{with probability } \frac{9}{19}. \end{cases}$$

Note that

$$\mu = E[W_i] = \frac{10}{19}(-100) + \frac{9}{19}100 = -\frac{100}{19};$$

$$\sigma = \text{std}(W_i) = E[(W_i)^2] - (E[W_i])^2 = \frac{600\sqrt{10}}{19}.$$

Let $W = \sum_{i=1}^{100} W_i$ be the total value of the winnings after betting 100 times. Since every bet is independent of any other bet, it follows that W is the sum of 100 independent identically distributed random variables. From the Central Limit Theorem, we find that

$$W \approx 100\mu + 10\sigma Z = -\frac{10000}{19} + \frac{6000\sqrt{10}}{19} Z.$$

The probability of winning at least \$1000 can be approximated as follows:

$$\begin{aligned} P(W > 1000) &\approx P\left(-\frac{10000}{19} + \frac{6000\sqrt{10}}{19} Z > 1000\right) \\ &= P(Z > 1.5284) = 0.0632. \end{aligned}$$

The probability of losing at least \$1000 can be approximated as follows:

$$\begin{aligned} P(W < -1000) &\approx P\left(-\frac{10000}{19} + \frac{6000\sqrt{10}}{19} Z < -1000\right) \\ &= P(Z < -0.4743) = 0.3176. \end{aligned}$$

We conclude that the probability of winning at least \$1000 is approximately 6%, and the probability of losing at least \$1000 is approximately 32%. □

Problem 9: Consider a put option with strike 55 and maturity 4 months on a non–dividend paying asset with spot price 60 which follows a lognormal model with drift $\mu = 0.1$ and volatility $\sigma = 0.3$. Assume that the risk-free rate is constant equal to 0.05.

(i) Find the probability that the put will expire in the money.

(ii) Find the risk–neutral probability that the put will expire in the money.

(iii) Compute $N(-d_2)$.

Solution: (i) The probability that the put option will expire in the money is equal to the probability that the spot price at maturity is lower than the strike price, i.e., to $P(S(T) < K)$. Recall that

$$\ln\left(\frac{S(T)}{S(0)}\right) = \left(\mu - q - \frac{\sigma^2}{2}\right)T + \sigma\sqrt{T}Z.$$

Then,

$$P(S(T) < K) = P\left(\frac{S(T)}{S(0)} < \frac{K}{S(0)}\right) = P\left(\ln\left(\frac{S(T)}{S(0)}\right) < \ln\left(\frac{K}{S(0)}\right)\right)$$

$$
\begin{aligned}
&= P\left(\left(\mu - q - \frac{\sigma^2}{2}\right)T + \sigma\sqrt{T}Z \;<\; \ln\left(\frac{K}{S(0)}\right)\right) \\
&= P\left(Z \;<\; \frac{\ln\left(\frac{K}{S(0)}\right) - \left(\mu - q - \frac{\sigma^2}{2}\right)T}{\sigma\sqrt{T}}\right) \\
&= N\left(\frac{\ln\left(\frac{K}{S(0)}\right) - \left(\mu - q - \frac{\sigma^2}{2}\right)T}{\sigma\sqrt{T}}\right).
\end{aligned}
$$

For $S = 60$, $K = 55$, $T = 1/3$, $\mu = 0.1$, $q = 0$, $\sigma = 0.3$, and $r = 0.05$, we obtain that the probability that the put will expire in the money is 0.271525, i.e., 27.1525%.

(ii) The risk–neutral probability that the put option will expire in the money is obtained just like the probability that the put expires in the money, by substituting the risk–free rate r for μ, i.e.,

$$
\begin{aligned}
P_{RN}(S(T) < K) &= P\left(Z \;<\; \frac{\ln\left(\frac{K}{S(0)}\right) - \left(r - q - \frac{\sigma^2}{2}\right)T}{\sigma\sqrt{T}}\right) \\
&= 0.304331 \;=\; 30.43\%. \qquad (4.30)
\end{aligned}
$$

(iii) Recall that

$$
d_2 \;=\; \frac{\ln\left(\frac{S(0)}{K}\right) + \left(r - q - \frac{\sigma^2}{2}\right)T}{\sigma\sqrt{T}}. \qquad (4.31)
$$

Then, $d_2 = 0.511983$, and

$$
N(-d_2) \;=\; 0.304331,
$$

which is the same as the risk–neutral probability that the put option will expire in the money; cf. (4.30).

To understand this result, note that

$$
P_{RN}(S(T) < K) \;=\; P\left(Z \;<\; -\frac{\ln\left(\frac{S(0)}{K}\right) + \left(r - q - \frac{\sigma^2}{2}\right)T}{\sigma\sqrt{T}}\right) \;=\; N(-d_2);
$$

cf. (4.30) and (4.31). □

Problem 10: (i) Consider an at-the-money call on a non–dividend paying asset; assume the Black-Scholes framework. Show that the Delta of the option is always greater than 0.5.

(ii) If the underlying asset pays dividends at the continuous rate q, when is the Delta of an at-the-money call less than 0.5?

Solution: (i) Recall that the Delta of a call option is given by

$$\Delta(C) = e^{-qT} N(d_1) = e^{-qT} N\left(\frac{\ln\left(\frac{S}{K}\right) + \left(r - q + \frac{\sigma^2}{2}\right) T}{\sigma\sqrt{T}}\right).$$

For an at–the–money call on a non–dividend paying asset, i.e., for $K = S$ and $q = 0$, we find that

$$\Delta(C) = N(d_1) = N\left(\frac{\left(r + \frac{\sigma^2}{2}\right)\sqrt{T}}{\sigma}\right) > N(0) = 0.5.$$

(ii) If the underlying asset pays dividends at the continuous rate q, the Delta of an ATM call is

$$\Delta(C) = e^{-qT} N(d_1) = e^{-qT} N\left(\frac{\left(r - q + \frac{\sigma^2}{2}\right)\sqrt{T}}{\sigma}\right).$$

For a fixed risk–free rate r and fixed maturity T, we conclude that $\Delta(C) < 0.5$ if and only if the dividend yield q and the volatility σ of the underlying asset satisfy the following condition:

$$N\left(\frac{\left(r - q + \frac{\sigma^2}{2}\right)\sqrt{T}}{\sigma}\right) < 0.5\, e^{qT}.$$

This happens, for example, if $r = q$ and T is large enough, since

$$\lim_{T\to\infty} N\left(\frac{\sigma\sqrt{T}}{2}\right) = 1 \quad \text{and} \quad \lim_{T\to\infty} 0.5\, e^{qT} = \infty. \quad \square$$

Problem 11: Consider a six months plain vanilla European put option with strike 50 on a lognormally distributed underlying asset paying dividends continuously at 2%. Assume that interest rates are constant at 4%.

Use risk–neutral valuation to write the value of the put as an integral over a finite interval. Find the value of the put option with six decimal digits accuracy using the Midpoint Rule and using Simpson's Rule. Also, compute the Black–Scholes value P_{BS} of the put and report the approximation errors of the numerical integration approximations at each step.

Solution: If the underlying asset follows a lognormal distribution, the value $S(T)$ of the underlying asset at maturity is a lognormal variable given by

$$\ln(S(T)) = \ln(S(0)) + \left(r - q - \frac{\sigma^2}{2}\right)T + \sigma\sqrt{T}Z,$$

where σ is the volatility of the underlying asset. Then, the probability density function $h(y)$ of $S(T)$ is

$$h(y) = \frac{1}{y\sigma\sqrt{2\pi T}} \exp\left(-\frac{\left(\ln y - \ln(S(0)) - \left(r - q - \frac{\sigma^2}{2}\right)T\right)^2}{2\sigma^2 T}\right), \quad (4.32)$$

if $y > 0$, and $h(y) = 0$ if $y \leq 0$.

Using risk–neutral valuation, we find that the value of the put is given by

$$\begin{aligned} P &= e^{-rT} E_{RN}[\max(K - S(T), 0)] \\ &= e^{-rT} \int_0^K (K - y)h(y)\, dy, \end{aligned} \quad (4.33)$$

where $h(y)$ is given by (4.32).

The Black–Scholes value of the put is $P_{BS} = 4.863603$. To compute a numerical approximation of the integral (4.33), we start with a partition of the interval $[0, K]$ into 4 intervals, and double the numbers of intervals up to 8192 intervals. We report the Midpoint Rule and Simpson's Rule approximations to (4.33) and the corresponding approximation errors to the Black–Scholes value P_{BS} in the table below:

No. Intervals	Midpoint Rule	Error	Simpson's Rule	Error
4	5.075832	0.212228	4.855908	0.007696
8	4.922961	0.059357	4.863955	0.000351
16	4.878220	0.014616	4.863631	0.000027
32	4.867248	0.003644	4.863611	0.000007
64	4.864518	0.000914	4.863610	0.000006
128	4.863837	0.000233	4.863610	0.000006
256	4.863666	0.000020	4.863609	0.000006
512	4.863624	0.000009	4.863609	0.000006
1024	4.863613	0.000006	4.863609	0.000006
2048	4.863610	0.000006	4.863609	0.000006
4096	4.863610	0.000006	4.863609	0.000006
8192	4.863610	0.000006	4.863609	0.000006

We first note that the approximation error does not go below $6 \cdot 10^{-6}$. This is due to the fact that the Black–Scholes value of the put, which is given by

$$P_{BS} = Ke^{-r(T-t)}N(-d_2) - Se^{-q(T-t)}N(-d_1),$$

is computed using numerical approximations to estimate the terms $N(-d_1)$ and $N(-d_2)$. The approximation error of these approximations is on the order of 10^{-7}. Using numerical integration, the real value of the put option is computed, but the error of the Black–Scholes value will propagate to the approximation errors of the numerical integration.

If we consider that convergence is achieved when the error is less than 10^{-5}, then convergence is achieved for 512 intervals for the Midpoint Rule and for 32 intervals for Simpson's Rule. This was to be expected given the quadratic convergence of the Midpoint Rule and the fourth order convergence of Simpson's Rule. □

Problem 12: Use risk–neutral pricing to price a supershare, i.e., an option that pays $(\max(S(T) - K, 0))^2$ at the maturity of the option. In other words, compute

$$V(0) = e^{-rT} E_{RN}[(\max(S(T) - K, 0))^2],$$

where the expected value is computed with respect to the risk–neutral distribution of the price $S(T)$ of the underlying asset at maturity T, which is assumed to follow a lognormal process with drift r and volatility σ. Assume that the underlying asset pays no dividends, i.e., $q = 0$.

Solution: Recall that

$$S(T) = S(0) \exp\left(\left(r - \frac{\sigma^2}{2}\right)T + \sigma\sqrt{T}Z\right)$$

and note that

$$S(T) \geq K \iff Z \geq \frac{\ln\left(\frac{K}{S(0)}\right) - \left(r - \frac{\sigma^2}{2}\right)T}{\sigma\sqrt{T}} = -d_2.$$

Then,

$$\begin{aligned} V(0) &= e^{-rT}\frac{1}{\sqrt{2\pi}}\int_{-d_2}^{\infty}\left(S(0)\exp\left(\left(r - \frac{\sigma^2}{2}\right)T + \sigma\sqrt{T}x\right) - K\right)^2 e^{-\frac{x^2}{2}}\,dx \\ &= K^2 e^{-rT}\frac{1}{\sqrt{2\pi}}\int_{-d_2}^{\infty} e^{-\frac{x^2}{2}}\,dx \\ &\quad - 2KS(0)\frac{e^{-rT}}{\sqrt{2\pi}}\int_{-d_2}^{\infty}\exp\left(\left(r - \frac{\sigma^2}{2}\right)T + \sigma\sqrt{T}x - \frac{x^2}{2}\right)\,dx \\ &\quad + S^2(0)\frac{e^{-rT}}{\sqrt{2\pi}}\int_{-d_2}^{\infty}\exp\left((2r - \sigma^2)T + 2\sigma\sqrt{T}x - \frac{x^2}{2}\right)\,dx. \qquad (4.34) \end{aligned}$$

When pricing a plain vanilla call using risk–neutrality, we proved that

$$\begin{aligned} C_{BS}(0) &= S(0)\frac{e^{-rT}}{\sqrt{2\pi}}\int_{-d_2}^{\infty}\exp\left(\left(r-\frac{\sigma^2}{2}\right)T+\sigma\sqrt{T}x-\frac{x^2}{2}\right)dx \\ &\quad - Ke^{-rT}\frac{1}{\sqrt{2\pi}}\int_{-d_2}^{\infty}e^{-\frac{x^2}{2}}dx \\ &= S(0)N(d_1) - Ke^{-rT}N(d_2). \end{aligned}$$

In other words, we showed that

$$e^{-rT}\frac{1}{\sqrt{2\pi}}\int_{-d_2}^{\infty}e^{-\frac{x^2}{2}}dx = e^{-rT}N(d_2); \qquad (4.35)$$

$$\frac{e^{-rT}}{\sqrt{2\pi}}\int_{-d_2}^{\infty}\exp\left(\left(r-\frac{\sigma^2}{2}\right)T+\sigma\sqrt{T}x-\frac{x^2}{2}\right)dx = N(d_1) \qquad (4.36)$$

From (4.34), (4.35), and (4.36), we conclude that

$$\begin{aligned} V(0) &= K^2e^{-rT}N(d_2) - 2KS(0)N(d_1) \\ &\quad +S^2(0)\frac{e^{-rT}}{\sqrt{2\pi}}\int_{-d_2}^{\infty}\exp\left((2r-\sigma^2)T+2\sigma\sqrt{T}x-\frac{x^2}{2}\right)dx. \end{aligned} \qquad (4.37)$$

The integral from (4.37) is computed by completing the square as follows:

$$\begin{aligned} & S^2(0)\frac{e^{-rT}}{\sqrt{2\pi}}\int_{-d_2}^{\infty}\exp\left((2r-\sigma^2)T+2\sigma\sqrt{T}x-\frac{x^2}{2}\right)dx \\ &= S^2(0)\frac{e^{-rT}}{\sqrt{2\pi}}\int_{-d_2}^{\infty}\exp\left(-\frac{(x-2\sigma\sqrt{T})^2}{2} + (2r+\sigma^2)T\right)dx \\ &= S^2(0)e^{(r+\sigma^2)T}\frac{1}{\sqrt{2\pi}}\int_{-d_2}^{\infty}\exp\left(-\frac{(x-2\sigma\sqrt{T})^2}{2}\right)dx \\ &= S^2(0)e^{(r+\sigma^2)T}\frac{1}{\sqrt{2\pi}}\int_{-(d_2+2\sigma\sqrt{T})}^{\infty}\exp\left(-\frac{y^2}{2}\right)dy \\ &= S^2(0)e^{(r+\sigma^2)T}N(d_2+2\sigma\sqrt{T}); \end{aligned} \qquad (4.38)$$

we used the substitution $y = x - 2\sigma\sqrt{T}$ and the fact that

$$\frac{1}{\sqrt{2\pi}}\int_{-a}^{\infty}\exp\left(-\frac{y^2}{2}\right)dy = \frac{1}{\sqrt{2\pi}}\int_{-\infty}^{a}\exp\left(-\frac{y^2}{2}\right)dy = N(a).$$

From (4.37) and (4.38), we conclude that

$$V(0) = K^2e^{-rT}N(d_2) - 2KS(0)N(d_1) + S^2(0)e^{(r+\sigma^2)T}N(d_2+2\sigma\sqrt{T}).$$

Problem 13: Use risk–neutral pricing to find the value of an option on a non–dividend–paying asset with lognormal distribution if the payoff of the option at maturity is equal to $\max((S(T))^\alpha - K, 0)$. Here, $\alpha > 0$ is a fixed constant.

Solution: Using risk–neutral pricing, we find that the value of the option is

$$V(0) = e^{-rT} E_{RN}[\max((S(T))^\alpha - K, 0)],$$

where

$$S(T) = S(0)\exp\left(\left(r - \frac{\sigma^2}{2}\right)T + \sigma\sqrt{T}Z\right). \tag{4.39}$$

Note that $(S(T))^\alpha \geq K$ is equivalent to $S(T) \geq K^{1/\alpha}$. Using (4.39), we find that

$$S(T) \geq K^{1/\alpha} \iff Z \geq \frac{\ln\left(\frac{K^{1/\alpha}}{S(0)}\right) - \left(r - \frac{\sigma^2}{2}\right) T}{\sigma\sqrt{T}} = -a.$$

Then,

$$V(0) = \frac{e^{-rT}}{\sqrt{2\pi}} \int_{-a}^{\infty} \left((S(0))^\alpha \exp\left(\alpha\left(r - \frac{\sigma^2}{2}\right)T + \alpha\sigma\sqrt{T}x\right) - K\right) e^{-\frac{x^2}{2}}\, dx.$$

Recall from (4.35) that

$$\frac{e^{-rT}}{\sqrt{2\pi}} \int_{-a}^{\infty} e^{-\frac{x^2}{2}} dx = e^{-rT} N(a).$$

Therefore,

$$\begin{aligned} V(0) = {} & \frac{(S(0))^\alpha}{\sqrt{2\pi}} \int_{-a}^{\infty} \exp\left((\alpha - 1)rT - \frac{\alpha\sigma^2}{2}T + \alpha\sigma\sqrt{T}x - \frac{x^2}{2}\right) dx \\ & - Ke^{-rT}N(a). \end{aligned}$$

By completing the square for the argument of the exponential function under the integral sign we obtain that

$$\begin{aligned} & \frac{1}{\sqrt{2\pi}} \int_{-a}^{\infty} \exp\left((\alpha - 1)rT - \frac{\alpha\sigma^2}{2}T + \alpha\sigma\sqrt{T}x - \frac{x^2}{2}\right) dx \\ = {} & \exp\left((\alpha - 1)rT - \frac{\alpha\sigma^2}{2}T + \frac{\alpha^2\sigma^2}{2}T\right) \cdot \\ & \frac{1}{\sqrt{2\pi}} \int_{-a}^{\infty} \exp\left(-\frac{(x - \alpha\sigma\sqrt{T})^2}{2}\right) dx \end{aligned}$$

$$= \exp\left((\alpha-1)\left(r+\frac{\alpha\sigma^2}{2}\right)T\right)\frac{1}{\sqrt{2\pi}}\int_{-(a+\alpha\sigma\sqrt{T})}^{\infty}\exp\left(-\frac{y^2}{2}\right)dy$$
$$= \exp\left((\alpha-1)\left(r+\frac{\alpha\sigma^2}{2}\right)T\right)\frac{1}{\sqrt{2\pi}}\int_{-\infty}^{a+\alpha\sigma\sqrt{T}}\exp\left(-\frac{y^2}{2}\right)dy$$
$$= \exp\left((\alpha-1)\left(r+\frac{\alpha\sigma^2}{2}\right)T\right)N(a+\alpha\sigma\sqrt{T});$$

note that the substitution $y = x - \alpha\sigma\sqrt{T}$ was used above.

We conclude that

$$\begin{aligned} V(0) &= (S(0))^{\alpha}\exp\left((\alpha-1)\left(r+\frac{\alpha\sigma^2}{2}\right)T\right)N(a+\alpha\sigma\sqrt{T}) \\ &\quad - Ke^{-rT}N(a), \end{aligned}$$

where

$$a = \frac{\ln\left(\frac{S(0)}{K^{1/\alpha}}\right)+\left(r-\frac{\sigma^2}{2}\right)T}{\sigma\sqrt{T}}. \quad \square$$

Problem 14: If the price of an asset follows a normal process, i.e., $dS = \mu dt + \sigma dX$, then

$$S(t_2) = S(t_1) + \mu(t_2-t_1) + \sigma\sqrt{t_2-t_1}\,Z, \quad \forall\, 0 < t_1 < t_2.$$

Assume that the risk free rate is constant and equal to r.

(i) Use risk neutrality to find the value of a call option with strike K and maturity T.

(ii) Use the Put–Call parity to find the value of a put option with strike K and maturity T, if the underlying asset follows a normal process as above.

Solution: (i) Using risk-neutral pricing, it follows that

$$C(0) = e^{-rT}E_{RN}[\max(S(T)-K,0)],$$

where the expected value is computed with respect to $S(T)$ given by

$$S(T) = S(0) + rT + \sigma\sqrt{T}\,Z. \tag{4.40}$$

Note that

$$S(T) > K \quad \text{iff} \quad Z > \frac{K-S(0)-rT}{\sigma\sqrt{T}} = d.$$

Then,

$$\begin{aligned} C(0) &= e^{-rT} \frac{1}{\sqrt{2\pi}} \int_d^\infty \left(S(0) + rT + \sigma\sqrt{T}x - K\right) e^{-\frac{x^2}{2}}\, dx \\ &= (S(0) + rT)e^{-rT} \frac{1}{\sqrt{2\pi}} \int_d^\infty e^{-\frac{x^2}{2}}\, dx \; - \; Ke^{-rT} \frac{1}{\sqrt{2\pi}} \int_d^\infty e^{-\frac{x^2}{2}}\, dx \\ &\quad + \frac{e^{-rT}\sigma\sqrt{T}}{\sqrt{2\pi}} \int_d^\infty xe^{-\frac{x^2}{2}} dx. \end{aligned}$$

Note that

$$\begin{aligned} \int_d^\infty xe^{-\frac{x^2}{2}} dx &= \lim_{t\to\infty} \int_d^t xe^{-\frac{x^2}{2}} dx = \lim_{t\to\infty} \left(-e^{-\frac{x^2}{2}}\right)\Big|_d^t = e^{-\frac{d^2}{2}}; \\ \frac{1}{\sqrt{2\pi}} \int_d^\infty e^{-\frac{x^2}{2}} dx &= 1 - \frac{1}{\sqrt{2\pi}} \int_{-\infty}^d e^{-\frac{x^2}{2}} dx = 1 - N(d) = N(-d), \end{aligned}$$

where $N(t)$ is the cumulative distribution of the standard normal variable.

We conclude that

$$\begin{aligned} C(0) &= (S(0) + rT)e^{-rT} N(-d) \; - \; Ke^{-rT} N(-d) \\ &\quad + \frac{e^{-rT}\sigma\sqrt{T}}{\sqrt{2\pi}} e^{-\frac{d^2}{2}}. \end{aligned} \qquad (4.41)$$

(ii) Regardless of the model used for describing the evolution of the price of the underlying asset, the Put–Call parity says that a portfolio made of a long position in a plain vanilla European call option and a short position in a plain vanilla European put option on the same asset and with the same strike and maturity as the call option has the same payoff at maturity as a long position in a unit of the underlying asset and a short cash position equal to the strike of the options. Using risk–neutral pricing, this can be written as

$$C(0) - P(0) = e^{-rT} E_{RN}[S(T) - K] = e^{-rT}(S(0) + rT - K), \qquad (4.42)$$

since $E_{RN}[S(T)] = S(0) + rT$; cf. (4.40).

From (4.41) and (4.42), we obtain that

$$\begin{aligned} P(0) &= C(0) \; - \; (S(0) + rT)e^{-rT} \; + \; Ke^{-rT} \\ &= Ke^{-rT}(1 - N(-d)) \; - \; (S(0) + rT)e^{-rT}(1 - N(-d)) \\ &\quad + \frac{e^{-rT}\sigma\sqrt{T}}{\sqrt{2\pi}} e^{-\frac{d^2}{2}} \\ &= Ke^{-rT} N(d) \; - \; (S(0) + rT)e^{-rT} N(d) \; + \; \frac{e^{-rT}\sigma\sqrt{T}}{\sqrt{2\pi}} e^{-\frac{d^2}{2}}, \end{aligned}$$

since $1 - N(-d) = N(d)$. □

Problem 15: (i) What is the value at time 0 of a derivative security on an underlying asset following a lognormal distribution with volatility σ which, at time T, pays $\frac{1}{S(T)}$, where $S(T)$ is the price of the underlying asset at time T? Assume that the risk free interest rates are constant and equal to r, and that the underlying asset pays dividends continuously at rate q.

(ii) Find the value at time 0 of a derivative security on the same underlying asset with the following payoff at maturity:

$$V(T) \;=\; \max\left(\frac{1}{K} - \frac{1}{S(T)},\, 0\right).$$

Solution: (i) Using risk–neutral pricing, we find that the value of the derivative security is

$$V(0) \;=\; e^{-rT}\, E_{RN}\left[\frac{1}{S(T)}\right],$$

where

$$S(T) \;=\; S(0)\exp\left(\left(r - q - \frac{\sigma^2}{2}\right)T + \sigma\sqrt{T}Z\right).$$

Therefore,

$$\begin{aligned}
V(0) &= \frac{e^{-rT}}{\sqrt{2\pi}}\int_{-\infty}^{\infty}\frac{1}{S(0)}\exp\left(-\left(r-q-\frac{\sigma^2}{2}\right)T - \sigma\sqrt{T}x\right)e^{-\frac{x^2}{2}}\,dx \\
&= \frac{\exp\left(-rT - \left(r-q-\frac{\sigma^2}{2}\right)T\right)}{S(0)}\cdot \\
&\quad \frac{1}{\sqrt{2\pi}}\int_{-\infty}^{\infty}\exp\left(-\sigma\sqrt{T}x - \frac{x^2}{2}\right)\,dx \\
&= \frac{\exp\left(\left(-2r+q+\frac{\sigma^2}{2}\right)T\right)}{S(0)}\cdot \\
&\quad \frac{1}{\sqrt{2\pi}}\int_{-\infty}^{\infty}\exp\left(\frac{\sigma^2}{2}T - \frac{1}{2}\left(x+\sigma\sqrt{T}\right)^2\right)\,dx \\
&= \frac{\exp\left((-2r+q+\sigma^2)T\right)}{S(0)}\cdot\frac{1}{\sqrt{2\pi}}\int_{-\infty}^{\infty}\exp\left(-\frac{1}{2}\left(x+\sigma\sqrt{T}\right)^2\right)\,dx.
\end{aligned}$$

Using the substitution $y = x + \sigma\sqrt{T}$ and the fact that the probability density function $\frac{1}{\sqrt{2\pi}}\exp\left(-\frac{y^2}{2}\right)$ integrates to 1 over the real axis, it is easy to see that

$$\frac{1}{\sqrt{2\pi}} \int_{-\infty}^{\infty} \exp\left(-\frac{1}{2}\left(x + \sigma\sqrt{T}\right)^2\right) dx = \frac{1}{\sqrt{2\pi}} \int_{-\infty}^{\infty} \exp\left(-\frac{y^2}{2}\right) dy = 1.$$

Thus, we conclude that the value of a derivative security paying $\frac{1}{S(T)}$ at time T is

$$\begin{aligned} V(0) &= \frac{\exp\left((-2r + q + \sigma^2)T\right)}{S(0)} \cdot \frac{1}{\sqrt{2\pi}} \int_{-\infty}^{\infty} \exp\left(-\frac{1}{2}\left(x + \sigma\sqrt{T}\right)^2\right) dx \\ &= \frac{1}{S(0)} \exp\left((-2r + q + \sigma^2)T\right). \end{aligned}$$

(ii) Recall that

$$S(T) = S(0) \exp\left(\left(r - q - \frac{\sigma^2}{2}\right)T + \sigma\sqrt{T}Z\right)$$

and note that

$$\begin{aligned} \frac{1}{K} \geq \frac{1}{S(T)} &\iff S(T) \geq K \\ &\iff Z \geq \frac{\ln\left(\frac{K}{S(0)}\right) - \left(r - q - \frac{\sigma^2}{2}\right)T}{\sigma\sqrt{T}} = d. \end{aligned}$$

Using risk–neutral pricing, we find that the value of the derivative security is

$$V(0) = e^{-rT} E_{RN}\left[\max\left(\frac{1}{K} - \frac{1}{S(T)}, 0\right)\right],$$

which can be written as

$$\begin{aligned} V(0) &= e^{-rT} \frac{1}{\sqrt{2\pi}} \cdot \\ &\quad \int_d^{\infty} \left(\frac{1}{K} - \frac{1}{S(0)} \exp\left(-\left(r - q - \frac{\sigma^2}{2}\right)T - \sigma\sqrt{T}x\right)\right) e^{-\frac{x^2}{2}} dx \\ &= \frac{e^{-rT}}{K} \cdot \frac{1}{\sqrt{2\pi}} \int_d^{\infty} e^{-\frac{x^2}{2}}\, dx - \frac{1}{S(0)} \cdot \qquad (4.43) \\ &\quad \frac{1}{\sqrt{2\pi}} \int_d^{\infty} \exp\left(\left(-2r + q + \frac{\sigma^2}{2}\right)T - \sigma\sqrt{T}x - \frac{x^2}{2}\right) dx. \qquad (4.44) \end{aligned}$$

From (4.35), it follows that

$$\frac{1}{\sqrt{2\pi}} \int_a^{\infty} e^{-\frac{x^2}{2}} dx = N(-a), \quad \forall\, a \in \mathbb{R}. \qquad (4.45)$$

By completing the square, it is easy to see that

$$\left(-2r+q+\frac{\sigma^2}{2}\right)T - \sigma\sqrt{T}x - \frac{x^2}{2} \;=\; (-2r+q+\sigma^2)T - \frac{(x+\sigma\sqrt{T})^2}{2}.$$

Thus,

$$\begin{aligned}
&\frac{1}{\sqrt{2\pi}}\int_d^\infty \exp\left(\left(-2r+q+\frac{\sigma^2}{2}\right)T - \sigma\sqrt{T}x - \frac{x^2}{2}\right)\,dx \\
&= \exp\left((-2r+q+\sigma^2)T\right)\cdot\frac{1}{\sqrt{2\pi}}\int_d^\infty \exp\left(-\frac{(x+\sigma\sqrt{T})^2}{2}\right)\,dx \\
&= \exp\left((-2r+q+\sigma^2)T\right)\cdot\frac{1}{\sqrt{2\pi}}\int_{d+\sigma\sqrt{T}}^\infty \exp\left(-\frac{y^2}{2}\right)\,dy; \qquad (4.46)
\end{aligned}$$

the substitution $y = x + \sigma\sqrt{T}$ was used above.

Using (4.45) for $a = d$ and for $a = d + \sigma\sqrt{T}$, we find that

$$\frac{1}{\sqrt{2\pi}}\int_d^\infty e^{-\frac{x^2}{2}}\,dx \;=\; N(-d); \qquad (4.47)$$

$$\frac{1}{\sqrt{2\pi}}\int_{d+\sigma\sqrt{T}}^\infty \exp\left(-\frac{y^2}{2}\right)\,dy \;=\; N\left(-d-\sigma\sqrt{T}\right) \qquad (4.48)$$

From (4.43), (4.44), (4.46), (4.47), and (4.48), we conclude that

$$V(0) \;=\; \frac{e^{-rT}}{K}N(-d) \;-\; \frac{1}{S(0)}\exp\left((-2r+q+\sigma^2)T\right)N\left(-d-\sigma\sqrt{T}\right). \quad \square$$

Chapter 5

Nonlinear solvers. N–dimensional Newton's method. Implied volatility. Bootstrapping.

5.1 Exercises

1. Use Newton's method to find the yield of a five year semiannual coupon bond with 3.375% coupon rate and price $100\,\frac{1}{32}$. What are the duration and convexity of the bond?

2. Compute the price, yield, modified duration, and convexity of a two year semiannual coupon bond with face value 100 and coupon rate 8%, if the zero rate curve is given by $r(0,t) = 0.05 + 0.01\ln\left(1+\frac{t}{2}\right)$.

3. Recall that finding the implied volatility from the given price of a call option is equivalent to solving the nonlinear problem $f(x) = 0$, where

$$f(x) \;=\; Se^{-qT}N(d_1(x)) - Ke^{-rT}N(d_2(x)) \;-\; C$$

and $d_1(x) \;=\; \dfrac{\ln\left(\frac{S}{K}\right) + \left(r-q+\frac{x^2}{2}\right)T}{x\sqrt{T}}$, $d_2(x) \;=\; \dfrac{\ln\left(\frac{S}{K}\right) + \left(r-q-\frac{x^2}{2}\right)T}{x\sqrt{T}}$.

(i) Show that $\lim_{x\to\infty} d_1(x) = \infty$ and $\lim_{x\to\infty} d_2(x) = -\infty$, and conclude that

$$\lim_{x\to\infty} f(x) \;=\; Se^{-qT} \;-\; C.$$

(ii) Show that

$$\lim_{x\searrow 0} d_1(x) \;=\; \lim_{x\searrow 0} d_2(x) \;=\; \begin{cases} -\infty, & \text{if } Se^{(r-q)T} < K; \\ 0, & \text{if } Se^{(r-q)T} = K; \\ \infty, & \text{if } Se^{(r-q)T} > K. \end{cases}$$

(Recall that $F = Se^{(r-q)T}$ is the forward price.)

Conclude that

$$\lim_{x \searrow 0} f(x) = \begin{cases} -C, & \text{if } Se^{(r-q)T} \leq K; \\ Se^{-qT} - Ke^{-rT} - C, & \text{if } Se^{(r-q)T} > K. \end{cases}$$

(iii) Show that $f(x)$ is a strictly increasing function and

$$\begin{array}{rcccl} -C & < & f(x) & < & Se^{-qT} - C, \text{ if } Se^{(r-q)T} \leq K; \\ Se^{-qT} - Ke^{-rT} - C & < & f(x) & < & Se^{-qT} - C, \text{ if } Se^{(r-q)T} > K. \end{array}$$

(iv) For what range of call option values does the problem $f(x) = 0$ have a positive solution? Compare your result to the range

$$Se^{-qT} - Ke^{-rT} < C < Se^{-qT}$$

required for obtaining a positive implied volatility for a value C of the call option.

4. A three months at–the–money call on an underlying asset with spot price 30 paying dividends continuously at a 2% rate is worth \$2.5. Assume that the risk free interest rate is constant at 6%.

   (i) Compute the implied volatility with six decimal digits accuracy, using the bisection method on the interval [0.0001, 1], the secant method with initial guess 0.5, and Newton's method with initial guess 0.5.

   (ii) Let σ_{imp} be the implied volatility previously computed using Newton's method. Use the formula

$$\sigma_{imp,approx} \approx \frac{\sqrt{2\pi}}{S\sqrt{T}} \frac{C - \frac{(r-q)T}{2}S}{1 - \frac{(r+q)T}{2}}.$$

   to compute an approximate value $\sigma_{imp,approx}$ for the implied volatility, and compute the relative error

$$\frac{|\sigma_{imp,approx} - \sigma_{imp}|}{\sigma_{imp}}.$$

5. Let $F : \mathbb{R}^3 \to \mathbb{R}^3$ given by

$$F(x) = \begin{pmatrix} x_1^3 + 2x_1x_2 + x_3^2 - x_2x_3 + 9 \\ 2x_1^2 + 2x_1x_2^2 + x_2^3x_3^2 - x_2^2x_3 - 2 \\ x_1x_2x_3 + x_1^3 - x_3^2 - x_1x_2^2 - 4 \end{pmatrix}.$$

The approximate gradient $\Delta_c F(x) = (\Delta_{c,j} F_i(x))_{i,j=1:n}$ of $F(x)$ is computed using central difference approximations, i.e.,

$$\Delta_{c,j} F_i(x) = \frac{F_i(x + he_j) - F_i(x - he_j)}{2h}, \quad j = 1 : n,$$

where e_j is a vector with all entries equal to 0 with the exception of the j-th entry, which is equal to 1.

(i) Solve $F(x) = 0$ using the approximate Newton's algorithm obtained by substituting $\Delta_c F(x_{old})$ for $\Delta F(x_{old})$. Use $h = 10^{-6}$, tol_consec $= 10^{-6}$, and tol_approx $= 10^{-9}$, and two different initial guesses: $x_0 = (1\ 2\ 3)^t$ and $x_0 = (2\ 2\ 2)^t$.

(ii) Compare these results to those corresponding to the approximate Newton's method with forward finite difference approximations for $\Delta F(x)$.

6. Consider a six months at–the–money call on an underlying asset following a lognormal distribution with volatility 30% and paying dividends continuously at rate q. Assume that the interest rates are constant at 4%. Show that there is a unique positive value of q such that $\Delta(C) = 0.5$, and find that value using Newton's method. How does this value of q compare to $r + \frac{\sigma^2}{2}$?

7. (i) Use bootstrapping to obtain a zero rate curve from the following prices of Treasury instruments with semiannual coupon payments:

	Coupon Rate	Price
3 − Month T-bill	0	98.7
6 − Month T-bill	0	97.5
2 − Year T-bond	4.875	$100\frac{5}{32}$
3 − Year T-bond	4.875	$100\frac{5}{32}$
5 − Year T-bond	4.625	$99\frac{22}{32}$
10 − Year T-bond	4.875	$101\frac{4}{32}$

Assume that interest is continuously compounded.

(ii) How would the zero rate curves obtained by bootstrapping from the bond prices above, one corresponding to semi-annually compounded interest, and the other one corresponding to continuously computed interest, compare? In other words, will one of the two curves be higher or lower than the other one, and why?

8. Use bootstrapping to obtain a continuously compounded zero rate curve given the prices of the following semiannual coupon bonds:

Maturity	Coupon Rate	Price
6 months	0	97.5
1 year	5	100
20 months	6	103
40 months	5	102
5 years	4	103

Assume that the overnight rate is 5% and that the zero rate curve is linear on the following time intervals:

$$[0, 0.5]; \quad [0.5, 1]; \quad \left[1, \frac{5}{3}\right]; \quad \left[\frac{5}{3}, \frac{10}{3}\right]; \quad \left[\frac{10}{3}, 5\right].$$

9. The following prices of the Treasury instruments are given:

	Coupon Rate	Price
6 – Month T-bill	0	99.4565
12 – Month T-bill	0	98.6196
2 – Year T-bond	2	$101\frac{17.5}{32}$
3 – Year T-bond	4.5	$107\frac{18}{32}$
5 – Year T-bond	3.125	$102\frac{8}{32}$
10 – Year T-bond	4	$103\frac{8.5}{32}$

The Treasury bonds pay semiannual coupons. Assume that interest is continuously compounded.

(i) Use bootstrapping to obtain a zero rate curve from the prices of the 6–months and 12–months Treasury bills, and of the 2–year, 5–year and 10–year Treasury bonds;

(ii) Find the relative pricing error corresponding to the 3–year Treasury bond if the zero rate curve obtained at part (i) is used. In other words, price a 3–year semiannual coupon bond with 4.5 coupon rate and find its relative error to the price $107\frac{18}{32}$ of the 3–year Treasury bond.

5.2 Solutions to Chapter 5 Exercises

Problem 1: Use Newton's method to find the yield of a five year semiannual coupon bond with 3.375% coupon rate and price $100\frac{1}{32}$. What are the duration and convexity of the bond?

Solution: Nine \$1.6875 coupon payments are made every six months, and a final payment of \$101.6875 is made after 5 years. By writing the value of the bond in terms of its yield, we obtain that

$$100 + \frac{1}{32} = \sum_{i=1}^{9} 1.6875 \exp\left(-y\,\frac{i}{2}\right) + 101.6875 \exp(-5y). \qquad (5.1)$$

We solve the nonlinear equation (5.1) for y using Newton's method. With initial guess $x_0 = 0.1$, Newton's method converges in four iterations to the solution $y = 0.033401$. We conclude that the yield of the bond is 3.3401%.

The duration and convexity of the bond are given by

$$D = \frac{1}{B}\left(\sum_{i=1}^{9} 1.6875\frac{i}{2} \exp\left(-y\,\frac{i}{2}\right) + 101.6875 \cdot 5 \exp(-5y)\right)$$

$$C = \frac{1}{B}\left(\sum_{i=1}^{9} 1.6875\frac{i^2}{4} \exp\left(-y\,\frac{i}{2}\right) + 101.6875 \cdot 25 \exp(-5y)\right),$$

where $B = 100 + \frac{1}{32}$ and $y = 0.033401$. We obtain that the duration of the bond is 4.642735 and the convexity of the bond is 22.573118. □

Problem 2: Compute the price, yield, modified duration, and convexity of a two year semiannual coupon bond with face value 100 and coupon rate 8%, if the zero rate curve is given by $r(0,t) = 0.05 + 0.01 \ln\left(1 + \frac{t}{2}\right)$.

Solution: The data below refers to the pseudocode from Table 2.5 of [1] for computing the price of a bond given the zero rate curve.
Input: $n = 4$; zero rate $r(0,t) = 0.05 + 0.01 \ln\left(1 + \frac{t}{2}\right)$;

$$\text{t\_cash\_flow} = [0.5\ 1\ 1.5\ 2]\,; \quad \text{v\_cash\_flow} = [4\ 4\ 4\ 104]\,.$$

Discount factors:

$$\text{disc} = [0.97422235\ \ 0.94738033\ \ 0.91998838\ \ 0.89238025].$$

Output: Bond price $B = 104.173911$.

In order to compute the duration and convexity of the bond, we first determine the yield of the bond. Using Newton's method, we obtain that the yield of the bond is 0.056792, i.e., 5.6792%. The modified duration and convexity of the bond are $D = 1.8901$ and $C = 3.6895$, respectively. □

Problem 3: Recall that finding the implied volatility from the given price of a call option is equivalent to solving the nonlinear problem $f(x) = 0$, where

$$f(x) \;=\; Se^{-qT}N(d_1(x)) - Ke^{-rT}N(d_2(x)) \;-\; C$$

and

$$d_1(x) = \frac{\ln\left(\frac{S}{K}\right) \;+\; \left(r - q + \frac{x^2}{2}\right)T}{x\sqrt{T}}; \quad d_2(x) = \frac{\ln\left(\frac{S}{K}\right) \;+\; \left(r - q - \frac{x^2}{2}\right)T}{x\sqrt{T}}.$$

(i) Show that $\lim_{x\to\infty} d_1(x) = \infty$ and $\lim_{x\to\infty} d_2(x) = -\infty$, and conclude that

$$\lim_{x\to\infty} f(x) \;=\; Se^{-qT} - C.$$

(ii) Show that

$$\lim_{x\searrow 0} d_1(x) \;=\; \lim_{x\searrow 0} d_2(x) \;=\; \begin{cases} -\infty, & \text{if } Se^{(r-q)T} < K; \\ 0, & \text{if } Se^{(r-q)T} = K; \\ \infty, & \text{if } Se^{(r-q)T} > K. \end{cases}$$

(Recall that $F = Se^{(r-q)T}$ is the forward price.)

Conclude that

$$\lim_{x\searrow 0} f(x) \;=\; \begin{cases} -C, & \text{if } Se^{(r-q)T} \le K; \\ Se^{-qT} - Ke^{-rT} \;-\; C, & \text{if } Se^{(r-q)T} > K. \end{cases}$$

(iii) Show that $f(x)$ is a strictly increasing function and

$$\begin{aligned} -C \;<\; f(x) \;&<\; Se^{-qT} - C, \;\text{ if } Se^{(r-q)T} \le K; \\ Se^{-qT} - Ke^{-rT} - C \;<\; f(x) \;&<\; Se^{-qT} - C, \;\text{ if } Se^{(r-q)T} > K. \end{aligned}$$

(iv) For what range of call option values does the problem $f(x) = 0$ have a positive solution? Compare your result to the range

$$Se^{-qT} - Ke^{-rT} \;<\; C \;<\; Se^{-qT}$$

required for obtaining a positive implied volatility for a value C of the call option.

Solution: (i) Recall that

$$d_1(x) \;=\; \frac{\ln\left(\frac{S}{K}\right) \;+\; (r-q)T}{x\sqrt{T}} \;+\; \frac{x\sqrt{T}}{2}; \tag{5.2}$$

$$d_2(x) \;=\; \frac{\ln\left(\frac{S}{K}\right) \;+\; (r-q)T}{x\sqrt{T}} \;-\; \frac{x\sqrt{T}}{2}. \tag{5.3}$$

It is easy to see that

$$\lim_{x\to\infty} d_1(x) = \infty \quad \text{and} \quad \lim_{x\to\infty} d_2(x) = -\infty,$$

and therefore

$$\lim_{x\to\infty} N(d_1(x)) = 1 \quad \text{and} \quad \lim_{x\to\infty} N(d_2(x)) = 0.$$

We conclude that

$$\lim_{x\to\infty} f(x) = Se^{-qT} - C.$$

(ii) Let $F = Se^{(r-q)T}$ be the forward price. From (5.2) and (5.3), it follows that $d_1(x)$ and $d_2(x)$ can be written as

$$d_1(x) = \frac{\ln\left(\frac{F}{K}\right)}{x\sqrt{T}} + \frac{x\sqrt{T}}{2}; \quad d_2(x) = \frac{\ln\left(\frac{F}{K}\right)}{x\sqrt{T}} - \frac{x\sqrt{T}}{2}.$$

- If $F < K$, then $\ln\left(\frac{F}{K}\right) < 0$ and

$$\lim_{x\searrow 0} d_1(x) = \lim_{x\searrow 0} d_2(x) = -\infty.$$

Therefore,

$$\lim_{x\searrow 0} N(d_1(x)) = \lim_{x\searrow 0} N(d_2(x)) = 0$$

and

$$\lim_{x\searrow 0} f(x) = -C.$$

- If $F = K$, then $d_1(x) = \frac{x\sqrt{T}}{2}$ and $d_2(x) = -\frac{x\sqrt{T}}{2}$, and therefore

$$\lim_{x\searrow 0} d_1(x) = \lim_{x\searrow 0} d_2(x) = 0.$$

Thus,

$$\lim_{x\searrow 0} N(d_1(x)) = \lim_{x\searrow 0} N(d_2(x)) = \frac{1}{2}$$

and

$$\begin{aligned}\lim_{x\searrow 0} f(x) &= \frac{1}{2}\left(Se^{-qT} - Ke^{-rT}\right) - C \\ &= \frac{e^{-rT}}{2}\left(Se^{(r-q)T} - K\right) - C \\ &= \frac{e^{-rT}}{2}(F - K) - C \\ &= -C.\end{aligned}$$

- If $F > K$, then $\ln\left(\frac{F}{K}\right) > 0$ and

$$\lim_{x \searrow 0} d_1(x) = \lim_{x \searrow 0} d_2(x) = \infty.$$

Therefore,

$$\lim_{x \searrow 0} N(d_1(x)) = \lim_{x \searrow 0} N(d_2(x)) = 1$$

and

$$\lim_{x \searrow 0} f(x) = Se^{-qT} - Ke^{-rT} - C.$$

(iii) Differentiating $f(x)$ with respect to x is the same as computing the derivative of the Black–Scholes value of a European call option with respect to the volatility σ, which is equal to the vega of the call. In other words,

$$f'(x) \;=\; \text{vega}(C) \;=\; Se^{-qT}\sqrt{T}\,\frac{1}{\sqrt{2\pi}}e^{-\frac{d_1^2}{2}},$$

where $d_1 = d_1(x) = \dfrac{\ln\left(\frac{S}{K}\right) + \left(r + \frac{x^2}{2}\right)T}{x\sqrt{T}}$.

Thus, $f'(x) > 0$, $\forall\, x > 0$, and $f(x)$ is strictly increasing.

Recall that $\lim_{x\to\infty} f(x) = Se^{-qT} - C$ and

$$\lim_{x \searrow 0} f(x) \;=\; \begin{cases} -C, & \text{if } F \le K; \\ Se^{-qT} - Ke^{-rT} - C, & \text{if } F > K. \end{cases}$$

Since $f(x)$ is strictly increasing, we conclude that

$$\begin{aligned} -C \;&<\; f(x) \;<\; Se^{-qT} - C, \quad \text{if } F \le K; \\ Se^{-qT} - Ke^{-rT} - C \;&<\; f(x) \;<\; Se^{-qT} - C, \quad \text{if } F > K. \end{aligned}$$

(iv) If $F \le K$, the problem $f(x) = 0$ has a solution $x > 0$ if and only if

$$0 \;<\; C \;<\; Se^{-qT}. \tag{5.4}$$

If $F > K$, the problem $f(x) = 0$ has a solution $x > 0$ if and only if

$$Se^{-qT} - Ke^{-rT} \;<\; C \;<\; Se^{-qT}. \tag{5.5}$$

Note that

$$Se^{-qT} - Ke^{-rT} \;=\; e^{-rT}(Se^{(r-q)T} - K) \;=\; e^{-rT}(F - K).$$

From (5.4) and (5.5), we conclude that the problem $f(x) = 0$ has a positive solution if and only if C belongs to the following range of values:

$$\max\left(Se^{-qT} - Ke^{-rT}, 0\right) \;<\; C \;<\; Se^{-qT}. \quad \square$$

Problem 4: A three months at–the–money call on an underlying asset with spot price 30 paying dividends continuously at a 2% rate is worth \$2.5. Assume that the risk free interest rate is constant at 6%.

(i) Compute the implied volatility with six decimal digits accuracy, using the bisection method on the interval $[0.0001, 1]$, the secant method with initial guess 0.5, and Newton's method with initial guess 0.5.

(ii) Let σ_{imp} be the implied volatility previously computed using Newton's method. Use the formula

$$\sigma_{imp,approx} \approx \frac{\sqrt{2\pi}}{S\sqrt{T}} \frac{C - \frac{(r-q)T}{2}S}{1 - \frac{(r+q)T}{2}}. \tag{5.6}$$

to compute an approximate value $\sigma_{imp,approx}$ for the implied volatility, and compute the relative error

$$\frac{|\sigma_{imp,approx} - \sigma_{imp}|}{\sigma_{imp}}.$$

Solution: (i) Both the secant method with $x_{-1} = 0.6$ and $x_0 = 0.5$ and Newton's method with initial guess $x_0 = 0.5$ converge in three iterations to an implied volatility of 39.7048%. The approximate values obtained at each iteration are given below:

k	Secant Method	Newton's Method
0	0.5	0.5
1	0.3969005134	0.3969152615
2	0.3970483533	0.3970481867
3	0.3970481868	0.3970481868

The bisection method on the interval $[0.0001, 1]$ converges in 30 iterations to the same implied volatility of 39.7048%. The first five iterations generate the following intervals:

$$\begin{array}{l} [0.250075, 0.5] \\ [0.375063, 0.5] \\ [0.375063, 0.437556] \\ [0.375063, 0.406309] \\ [0.390686, 0.406309] \end{array}$$

(ii) The approximate value for the implied volatility given by (5.6) is

$$\sigma_{imp,approx} = 0.3966718145 = 39.6672\%.$$

If $\sigma_{imp} = 0.3970481868$ is the implied volatility obtained using Newton's method, then

$$\frac{|\sigma_{imp,approx} - \sigma_{imp}|}{\sigma_{imp}} = 0.000948 = 0.0948\%. \quad \square$$

Problem 5: Let $F : \mathbb{R}^3 \to \mathbb{R}^3$ given by

$$F(x) = \begin{pmatrix} x_1^3 + 2x_1x_2 + x_3^2 - x_2x_3 + 9 \\ 2x_1^2 + 2x_1x_2^2 + x_2^3x_3^2 - x_2^2x_3 - 2 \\ x_1x_2x_3 + x_1^3 - x_3^2 - x_1x_2^2 - 4 \end{pmatrix}.$$

The approximate gradient $\Delta_c F(x) = (\Delta_{c,j} F_i(x))_{i,j=1:n}$ of $F(x)$ is computed using central difference approximations, i.e.,

$$\Delta_{c,j} F_i(x) = \frac{F_i(x + he_j) - F_i(x - he_j)}{2h}, \quad j = 1:n,$$

where e_j is a vector with all entries equal to 0 with the exception of the j-th entry, which is equal to 1.

(i) Solve $F(x) = 0$ using the approximate Newton's algorithm obtained by substituting $\Delta_c F(x_{old})$ for $\Delta F(x_{old})$. Use $h = 10^{-6}$, tol_consec $= 10^{-6}$, and tol_approx $= 10^{-9}$, and two different initial guesses:

$$x_0 = \begin{pmatrix} 1 \\ 2 \\ 3 \end{pmatrix}$$

and

$$x_0 = \begin{pmatrix} 2 \\ 2 \\ 2 \end{pmatrix}.$$

(ii) Compare these results to those corresponding to the approximate Newton's method with forward finite difference approximations for $\Delta F(x)$.

Solution: We use Newton's method and the approximate Newton's method both with forward difference approximations and with central difference approximations with tol_consec $= 10^{-6}$ and tol_approx $= 10^{-9}$. The parameter h is chosen to be equal to tol_consec, i.e., $h = 10^{-6}$.

Both algorithms converged to the same solutions, as follows:

$$x^* = \begin{pmatrix} -1.6905507599 \\ 1.9831072429 \\ -0.8845580785 \end{pmatrix} \quad \text{for} \quad x_0 = \begin{pmatrix} 1 \\ 2 \\ 3 \end{pmatrix}$$

and

$$x^* = \begin{pmatrix} -1 \\ 3 \\ 1 \end{pmatrix} \quad \text{for} \quad x_0 = \begin{pmatrix} 2 \\ 2 \\ 2 \end{pmatrix}.$$

The iteration counts are given in the table below:

	$x_0 = \begin{pmatrix} 1 \\ 2 \\ 3 \end{pmatrix}$	$x_0 = \begin{pmatrix} 2 \\ 2 \\ 2 \end{pmatrix}$
Iteration Count Newton's Method	9	40
Iteration Count Approximate Newton Forward Differences	9	65
Iteration Count Approximate Newton Central Differences	9	43

We note that using central finite differences approximates the gradient $DF(x)$ of $F(x)$ more accurately than if forward finite differences are used, resulting in algorithms with iteration counts closer to the iteration counts for Newton's method. □

Problem 6: Consider a six months at–the–money call on an underlying asset following a lognormal distribution with volatility 30% and paying dividends continuously at rate q. Assume that the interest rates are constant at 4%. Show that there is a unique positive value of q such that $\Delta(C) = 0.5$, and find that value using Newton's method. How does this value of q compare to $r + \frac{\sigma^2}{2}$?

Solution: Recall that the Delta of a plain vanilla call option on an underlying asset paying dividends continuously at rate q is

$$\Delta(C) = e^{-qT} N(d_1), \tag{5.7}$$

where

$$d_1 = \frac{\ln\left(\frac{S}{K}\right) + \left(r - q + \frac{\sigma^2}{2}\right) T}{\sigma\sqrt{T}}.$$

For an at–the–money option, i.e., for $S = K$, we obtain that

$$d_1 = \frac{(r-q)\sqrt{T}}{\sigma} + \frac{\sigma\sqrt{T}}{2}. \tag{5.8}$$

From (5.7) and (5.8), we find that

$$\Delta(C) = e^{-qT} N(d_1) = e^{-qT} N\left(\frac{(r-q)\sqrt{T}}{\sigma} + \frac{\sigma\sqrt{T}}{2}\right). \qquad (5.9)$$

It is easy to see that $\Delta(C)$ is a decreasing function of the dividend rate q, since

$$\begin{aligned}\frac{\partial \Delta}{\partial q} &= -Te^{-qT}N(d_1) + e^{-qT}N'(d_1)\frac{\partial d_1}{\partial q} \\ &= -Te^{-qT}N(d_1) + e^{-qT}\frac{1}{\sqrt{2\pi}}e^{-\frac{d_1^2}{2}}\left(-\frac{\sqrt{T}}{\sigma}\right) \\ &< 0.\end{aligned}$$

When $q = 0$, we find that

$$\Delta(C) = N\left(\frac{r\sqrt{T}}{\sigma} + \frac{\sigma\sqrt{T}}{2}\right) > N(0) = 0.5.$$

Also, since $0 < N(d_1) < 1$, it follows that

$$\lim_{q\to\infty} \Delta(C) = \lim_{q\to\infty} \left(e^{-qT}N(d_1)\right) = 0.$$

We conclude that $\Delta(C)$ is a decreasing function of q and that, for $q \geq 0$, the values of $\Delta(C)$ decrease from $N(0)$ to 0. Since $N(0) > 0.5$, there exists a unique value $q > 0$ such that $\Delta(C) = 0.5$. From (5.9), it follows that this value can be obtained by solving for x the nonlinear equation $f(x) = 0$, where

$$f(x) = e^{-xT} N\left(\frac{(r-x)\sqrt{T}}{\sigma} + \frac{\sigma\sqrt{T}}{2}\right) - 0.5.$$

Using Newton's method, we obtain that $q = x = 0.066906$.

In other words, if the interest rates are 4%, the Delta of a six months at–the–money call option on an underlying asset with volatility 30% is equal to 0.5 if the underlying asset pays 6.69% dividends continuously.

If $q = r + \frac{\sigma^2}{2}$, we find from (5.8) and (5.9) that $d_1 = 0$ and

$$\Delta(C) = e^{-qT}N(0) = 0.5e^{-qT} < 0.5.$$

Since $\Delta(C)$ is a decreasing function of q, we obtain that the value of q such that $\Delta(C) = 0.5$ must be lower than $r + \frac{\sigma^2}{2}$. Indeed, the value previously obtained for q satisfies this condition, i.e.,

$$q = 0.066906 < 0.085 = r + \frac{\sigma^2}{2}. \quad \square$$

Problem 7: (i) Use bootstrapping to obtain a zero rate curve from the following prices of Treasury instruments with semiannual coupon payments:

	Coupon Rate	Price
3 – Month T-bill	0	98.7
6 – Month T-bill	0	97.5
2 – Year T-bond	4.875	$100\frac{5}{32}$
3 – Year T-bond	4.875	$100\frac{5}{32}$
5 – Year T-bond	4.625	$99\frac{22}{32}$
10 – Year T-bond	4.875	$101\frac{4}{32}$

Assume that interest is continuously compounded.

(ii) How would the zero rate curves obtained by bootstrapping from the bond prices above, one corresponding to semi-annually compounded interest, and the other one corresponding to continuously computed interest, compare? In other words, will one of the two curves be higher or lower than the other one, and why?

Solution: (i) For the Treasury bills, the zero rates can be computed directly:

$$\begin{aligned} r(0,0.25) &= 4\ln\left(\frac{100}{98.7}\right) = 0.052341 = 5.2341\%; \\ r(0,0.5) &= 2\ln\left(\frac{100}{97.5}\right) = 0.050636 = 5.0636\%. \end{aligned}$$

Bootstrapping is needed to obtain the 2–year, 3–year, 5–year and 10–year zero rates.

For example, for the two year bond, if the zero rate curve is assumed to be linear between six months and two years, then

$$r(0,t) = \frac{(2-t)r(0,0.5)+(t-0.5)r(0,2)}{1.5}, \quad \forall\, t \in [0.5, 2]. \qquad (5.10)$$

If we let $x = r(0,2)$, we find from (5.10) that

$$r(0,1) = \frac{r(0,0.5)+0.5x}{1.5}; \quad r(0,1.5) = \frac{0.5r(0,0.5)+x}{1.5}.$$

Recall that the price of the two year bond is the discounted present value of all the future cash flows of the bond. Then,

$$\begin{aligned} 100+\frac{5}{32} &= 2.4375\, e^{-0.5r(0,0.5)} + 2.4375\, e^{-r(0,1)} \\ &\quad + 2.4375\, e^{-1.5r(0,1.5)} + 102.4375\, e^{-2r(0,2)} \\ &= 2.4375\, e^{-0.5r(0,0.5)} + 2.4375 \exp\left(-\frac{r(0,0.5)+0.5x}{1.5}\right) \\ &\quad +2.4375\exp\left(-1.5\,\frac{0.5r(0,0.5)+x}{1.5}\right) + 102.4375\, e^{-2x}. \end{aligned}$$

Using Newton's method to solve the nonlinear equation above for x, we obtain that $x = 0.047289$, and therefore

$$r(0,2) \;=\; 4.7289\%.$$

We proceed by assuming that the zero rate curve is linear between two years and three years. We note that $r(0,0.5)$, $r(0,1)$, $r(0,1.5)$, and $r(0,2)$ are known. If we let $x = r(0,3)$, the price of the three year bond can be written as

$$\begin{aligned} 100 + \frac{5}{32} \;=\; & 2.4375\ e^{-0.5r(0,0.5)} \;+\; 2.4375\ e^{-r(0,1)} \;+\; 2.4375\ e^{-1.5r(0,1.5)} \\ & + 2.4375\ e^{-2r(0,2)} \;+\; 2.4375\ \exp\left(-2.5\ \frac{r(0,2)+x}{2}\right) \\ & + 102.4375\ \exp\left(-2x\right) \end{aligned}$$

Using Newton's method to solve the nonlinear problem above, we obtain that $x = 0.047582$. Therefore

$$r(0,3) \;=\; 4.7582\%.$$

Using bootstrapping, we obtain similarly that

$$r(0,5) \;=\; 4.6303\% \quad \text{and} \quad r(0,10) \;=\; 4.6772\%.$$

Thus, we found the following zero rates corresponding to the maturities of the four given bonds, i.e., 3 months, 6 months, 2 years, 3 years, 5 years, and 10 years:

$$\begin{aligned} r(0,0.25) &= 5.2341\%; \quad r(0,0.5) = 5.0636\%; \quad r(0,2) = 4.7289\%; \\ r(0,3) &= 4.7582\%; \quad r(0,5) = 4.6303\%; \quad r(0,10) = 4.6772\%. \end{aligned}$$

Since we assumed that the zero rate curve is linear between any two consecutive bond maturities, the zero rate $r(0,t)$ is known for any time between the shortest and longest bond maturities, i.e., for any $t \in [0.25, 10]$.

(ii) Denote by $r_c(0,t)$ and $r_2(0,t)$ the zero rate curves corresponding to identical discount factors, with $r_c(0,t)$ corresponding to continuously compounded interest, and with $r_2(0,t)$ corresponding to semi-annually compounded interest. Then,

$$e^{-tr_c(0,t)} \;=\; \left(1 + \frac{r_2(0,t)}{2}\right)^{-2t}, \quad \forall\, t > 0. \tag{5.11}$$

By solving for $r_c(0,t)$ in (5.11), we find that

$$r_c(0,t) \;=\; \ln\left(\left(1 + \frac{r_2(0,t)}{2}\right)^2\right)$$

$$\begin{aligned}
&= r_2(0,t)\ \ln\left(\left(1+\frac{r_2(0,t)}{2}\right)^{\frac{2}{r_2(0,t)}}\right) \\
&= r_2(0,t)\ \ln\left(\left(1+\frac{1}{2/r_2(0,t)}\right)^{2/r_2(0,t)}\right) \\
&< r_2(0,t);
\end{aligned}$$

for the last inequality we used the fact that

$$\left(1+\frac{1}{x}\right)^x \ < \ e, \quad \forall x > 0,$$

for $x = 2/r_2(0,t)$. In other words, the semi-annually compounded zero rate curve is higher than the continuously compounded zero rate curve if both curves have the same discount factors.

While a rigorous proof is much more technical, the same happens if the two curves are obtained by bootstrapping from the same set of bonds, i.e., the zero rates corresponding to each bond maturity are higher if interest is compounded semi-annually than if interest is compounded continuously. This is done sequentially, beginning with the zero rates corresponding to the shortest bond maturity and moving to the zero rates corresponding to the longest bond maturity one bond maturity at a time. □

Problem 8: Use bootstrapping to obtain a continuously compounded zero rate curve given the prices of the following semiannual coupon bonds:

Maturity	Coupon Rate	Price
6 months	0	97.5
1 year	5	100
20 months	6	103
40 months	5	102
5 years	4	103

Assume that the overnight rate is 5% and that the zero rate curve is linear on the following time intervals:

$$[0, 0.5]; \quad [0.5, 1]; \quad \left[1, \frac{5}{3}\right]; \quad \left[\frac{5}{3}, \frac{10}{3}\right]; \quad \left[\frac{10}{3}, 5\right].$$

Solution: We know that $r(0,0) = 0.05$. The six months zero rate can be computed from the price of the 6–months zero coupon bond as

$$r(0,0.5) \ = \ 2\ln\left(\frac{100}{97.5}\right) = 0.050636 = 5.0636\%. \tag{5.12}$$

Using (5.12), we can solve for the zero rate $r(0,1)$ from the formula given the price of the one year bond, i.e.,

$$100 \;=\; 2.5\; e^{-0.5r(0,0.5)} \;+\; 102.5\; e^{-r(0,1)}$$

and obtain that

$$r(0,1) \;=\; 0.049370 = 4.9370\%.$$

The third bond pays coupons in 2, 8, 14, and 20 months, when it also pays the face value of the bond. Then,

$$\begin{aligned} 103 \;=\;& 3\exp\left(-\frac{2}{12}r\left(0,\frac{2}{12}\right)\right) \;+\; 3\exp\left(-\frac{8}{12}r\left(0,\frac{8}{12}\right)\right) \\ &+ 3\exp\left(-\frac{14}{12}r\left(0,\frac{14}{12}\right)\right) + \; 103\exp\left(-\frac{20}{12}r\left(0,\frac{20}{12}\right)\right). \end{aligned} \tag{5.13}$$

Since we assumed that the zero rate curve is linear on the intervals $[0,0.5]$ and $[0.5,1]$, the zero rates $r(0,\frac{2}{12})$ and $r(0,\frac{8}{12})$ are known and can be obtained by linear interpolation as follows:

$$r\left(0,\frac{2}{12}\right) \;=\; \frac{4r(0,0)+2r(0,0.5)}{6} \;=\; 0.050212; \tag{5.14}$$

$$r\left(0,\frac{8}{12}\right) \;=\; \frac{4r(0,0.5)+2r(0,1)}{6} \;=\; 0.050214. \tag{5.15}$$

Let $x = r\left(0,\frac{20}{12}\right)$. Since $r(0,t)$ is linear on the interval $\left[1,\frac{20}{12}\right]$, we find that

$$r\left(0,\frac{14}{12}\right) \;=\; \frac{6r(0,1)+2x}{8}. \tag{5.16}$$

From (5.16), it follows that the formula (5.13) can be written as

$$\begin{aligned} 103 \;=\;& 3\exp\left(-\frac{1}{6}r\left(0,\frac{2}{12}\right)\right) \;+\; 3\exp\left(-\frac{2}{3}r\left(0,\frac{8}{12}\right)\right) \\ &+ 3\exp\left(-\frac{7}{6}\cdot\frac{6r(0,1)+2x}{8}\right) \;+\; 103\exp\left(-\frac{5}{3}x\right), \end{aligned} \tag{5.17}$$

where $r\left(0,\frac{2}{12}\right)$ and $r\left(0,\frac{8}{12}\right)$ are given by (5.14) and (5.15), respectively. Using Newton's method to solve for x in (5.17), we find that $x = 0.052983$, and therefore

$$r\left(0,\frac{20}{12}\right) \;=\; 5.2983\%.$$

Bootstrapping for the fourth and fifth bonds proceed similarly. For example, the fourth bond makes coupon payments in 4, 10, 16, 22, 28, 34, and 40 months. The zero rates corresponding to coupon dates less than 20 months, i.e., to the coupon dates 4, 10 and 16 months, can be obtained from the part of the zero curve that was already determined. By setting $x = r\left(0, \frac{40}{12}\right)$ and assuming that the zero rate curve is linear between 20 months and 40 months, the zero rates corresponding to 22, 28, 34, and 40 months can be written in terms of x. Thus, the pricing formula for the fourth bond becomes a nonlinear equation in x which can be solved using Newton's method. The zero rate $r\left(0, \frac{40}{12}\right)$ is then determined.

Using bootstrapping and Newton's method we obtain that

$$r\left(0, \frac{40}{12}\right) = 4.5326\%; \quad r(0,5) = 3.2119\%.$$

Summarizing, the zero rate curve obtained by bootstrapping is given by

$$r(0,0) = 0.05; \quad r\left(0, \frac{2}{12}\right) = 0.050212; \quad r\left(0, \frac{8}{12}\right) = 0.050214;$$

$$r\left(0, \frac{20}{12}\right) = 0.052983; \quad r\left(0, \frac{40}{12}\right) = 0.045326; \quad r(0,5) = 0.032119,$$

and is linear on the intervals

$$[0, 0.5]; \quad [0.5, 1]; \quad \left[1, \frac{5}{3}\right]; \quad \left[\frac{5}{3}, \frac{10}{3}\right]; \quad \left[\frac{10}{3}, 5\right]. \quad \square$$

Problem 9: The following prices of the Treasury instruments are given:

	Coupon Rate	Price
6 – Month T-bill	0	99.4565
12 – Month T-bill	0	98.6196
2 – Year T-bond	2	$101\frac{17.5}{32}$
3 – Year T-bond	4.5	$107\frac{18}{32}$
5 – Year T-bond	3.125	$102\frac{8}{32}$
10 – Year T-bond	4	$103\frac{8.5}{32}$

The Treasury bonds pay semiannual coupons. Assume that interest is continuously compounded.

(i) Use bootstrapping to obtain a zero rate curve from the prices of the 6–months and 12–months Treasury bills, and of the 2–year, 5–year and 10–year Treasury bonds;

(ii) Find the relative pricing error corresponding to the 3–year Treasury bond if the zero rate curve obtained at part (i) is used. In other words, price a 3–year semiannual coupon bond with 4.5 coupon rate and find its relative error to the price $107\frac{18}{32}$ of the 3–year Treasury bond.

Solution: (i) The 6–months and 12–months zero rates can be obtained directly from the prices of the Treasury bills, i.e.,

$$\begin{aligned} r(0,0.5) &= 2\ln\left(\frac{100}{99.4565}\right) = 1.09\%; \\ r(0,1) &= \ln\left(\frac{100}{98.6196}\right) = 1.39\%. \end{aligned}$$

Using bootstrapping, we obtain the following 2–year, 5–year and 10–year zero rates:

$$\begin{aligned} r(0,2) &= 1.2099\%; \\ r(0,5) &= 2.6824\%; \\ r(0,10) &= 3.7371\%. \end{aligned}$$

(ii) The zero rates computed above correspond to the following zero rate curve which is piecewise linear between consecutive bond maturities:

$$r(0,t) = \begin{cases} (2t-1)\, r(0,1) + 2(1-t)\, r(0,0.5), & \text{if } 0.5 \le t \le 1; \\ (t-1)\, r(0,2) + (2-t)\, r(0,1), & \text{if } 1 \le t \le 2; \\ \frac{t-2}{3}\, r(0,5) + \frac{5-t}{3}\, r(0,2), & \text{if } 2 \le t \le 5; \\ \frac{t-5}{5}\, r(0,10) + \frac{10-t}{5}\, r(0,5), & \text{if } 5 \le t \le 10. \end{cases}$$

With this zero rate curve, the value of the 3–year semiannual coupon bond with 4.5 coupon rate is

$$\begin{aligned} B &= \sum_{i=1}^{5} 2.25 \exp\left(-r\left(0, \frac{i}{2}\right)\frac{i}{2}\right) + 102.25 \exp(-3r(0,3)) \\ &= 108.1930. \end{aligned}$$

The price of the 3–year Treasury bond was given to be 107.5625. Therefore, the relative pricing error given by the bootstrapped zero rate curve which does not include the 3–year bond is

$$\frac{|107.5625 - 108.1930|}{107.5625} = 0.005862 = 0.59\%. \quad \square$$

Chapter 6

Taylor's formula and Taylor series. ATM approximation of Black–Scholes formulas.

6.1 Exercises

1. Show that the cubic Taylor approximation of $\sqrt{1+x}$ around 0 is

$$\sqrt{1+x} \approx 1 + \frac{x}{2} - \frac{x^2}{8} + \frac{x^3}{16}.$$

2. Use the Taylor series expansion of the function e^x to find the value of $e^{0.25}$ with six decimal digits accuracy.

3. Show that

$$e^{-x} - \frac{1}{1+x} = O(x^2), \quad \text{as } x \to 0.$$

4. (i) Let $g(x)$ be an infinitely differentiable function. Find the linear and quadratic Taylor approximations of $e^{g(x)}$ around the point 0.

   (ii) Use the result above to compute the quadratic Taylor approximation around 0 of $e^{(x+1)^2}$.

   (iii) Compute the quadratic Taylor approximation around 0 of $e^{(x+1)^2}$ by using Taylor approximations of e^x and e^{x^2}.

5. Find the Taylor series expansion of the functions

$$\ln(1-x^2) \quad \text{and} \quad \frac{1}{1-x^2}$$

   around the point 0, using the Taylor series expansions of

$$\ln(1-x) \quad \text{and} \quad \frac{1}{1-x}$$

around 0.

6. Recall that

$$\left(1+\frac{1}{x}\right)^{x} \; < \; e \; < \; \left(1+\frac{1}{x}\right)^{x+1}, \quad \forall\, x \geq 1.$$

Prove that

$$\left(1+\frac{1}{x}\right)^{x+\frac{1}{2}-\frac{1}{12x}} \; < \; e \; < \; \left(1+\frac{1}{x}\right)^{x+\frac{1}{2}}, \quad \forall\, x \geq 1.$$

7. Compute the Taylor series expansion of

$$\ln\left(\frac{1+x}{1-x}\right)$$

around the point 0, and find its radius of convergence.

8. (i) Find the radius of convergence of the series

$$1 + \frac{x^4}{2!} + \frac{x^8}{4!} + \frac{x^{12}}{6!} + \ldots \tag{6.1}$$

(ii) Show that the series from (6.1) is the Taylor series expansion of the function

$$\frac{e^{x^2} + e^{-x^2}}{2}.$$

9. Let

$$T(x) \; = \; \sum_{k=1}^{\infty} \frac{(-1)^{k+1}\, x^k}{k}$$

be the Taylor series expansion of $f(x) = \ln(1+x)$. Our goal is to show that $T(x) = f(x)$ for all x such that $|x| < 1$.

Let

$$P_n(x) \; = \; \sum_{k=1}^{n} \frac{(-1)^{k+1}\, x^k}{k}$$

be the Taylor polynomial of degree n corresponding to $f(x)$. Since $T(x) = \lim_{n\to\infty} P_n(x)$, it follows that $f(x) = T(x)$ for all $|x| < 1$ if and only if

$$\lim_{n\to\infty} |f(x) - P_n(x)| \; = \; 0, \quad \forall\, |x| < 1. \tag{6.2}$$

(i) Show that, for any x,

$$f(x) - P_n(x) = \int_0^x \frac{(-1)^{n+2}\,(x-t)^n}{(1+t)^{n+1}}\,dt.$$

(ii) Show that, for any $0 \le x < 1$,

$$|f(x) - P_n(x)| \le \int_0^x \left(\frac{x-t}{1+t}\right)^n \frac{1}{1+t}\,dt \le x^n \ln(1+x). \qquad (6.3)$$

Use (6.3) to prove that (6.2) holds for all x such that $0 \le x < 1$.

(iii) Assume that $-1 < x \le 0$. Let $s = -x$. Show that

$$|f(x) - P_n(x)| = \int_0^s \frac{(s-z)^n}{(1-z)^{n+1}}\,dz.$$

Note that $\frac{s-z}{1-z} \le s$, for all $0 \le z \le s < 1$, and obtain that

$$|f(x) - P_n(x)| \le s^n |\ln(1-s)| = (-x)^n |\ln(1+x)|.$$

Conclude that (6.2) holds true for all x such that $-1 < x \le 0$.

10. The goal of this exercise is to compute

$$\int_0^1 \ln(1-x)\ln(x)\,dx. \qquad (6.4)$$

(i) Show that

$$\lim_{x \searrow 0} (\ln(1-x)\ln(x)) = \lim_{x \nearrow 1} (\ln(1-x)\ln(x)) = 0,$$

and conclude that the integral (6.4) can be regarded as a definite integral.

(ii) Use the Taylor series expansion of $\ln(1-x)$ for $|x| < 1$ to show that

$$\int_0^1 \ln(1-x)\ln(x)\,dx = -\sum_{n=1}^{\infty} \frac{1}{n} \int_0^1 x^n \ln(x)\,dx.$$

(iii) Prove that

$$\int_0^1 \ln(1-x)\ln(x)\,dx = \sum_{k=1}^{\infty} \frac{1}{n(n+1)^2}.$$

(iv) Use that fact that

$$\sum_{k=1}^{\infty} \frac{1}{n^2} = \frac{\pi^2}{6}$$

to obtain that

$$\int_0^1 \ln(1-x)\ln(x)\,dx = 2 - \frac{\pi^2}{6}.$$

11. In the Cox-Ross-Rubinstein parametrization for a binomial tree, the up and down factors u and d, and the risk–neutral probability p of the price going up during one time step are

$$u = A + \sqrt{A^2 - 1}; \tag{6.5}$$

$$d = A - \sqrt{A^2 - 1}; \tag{6.6}$$

$$p = \frac{e^{r\delta t} - d}{u - d}, \tag{6.7}$$

where

$$A = \frac{1}{2}\left(e^{-r\delta t} + e^{(r+\sigma^2)\delta t}\right).$$

Use Taylor expansions to show that, for a small time step δt, u, d and p may be approximated by

$$u = e^{\sigma\sqrt{\delta t}}; \tag{6.8}$$

$$d = e^{-\sigma\sqrt{\delta t}}; \tag{6.9}$$

$$p = \frac{1}{2} + \frac{1}{2}\left(\frac{r}{\sigma} - \frac{\sigma}{2}\right)\sqrt{\delta t}. \tag{6.10}$$

In other words, write the Taylor expansions for (6.5–6.7) and for (6.8–6.10) and show that they are identical if all the terms of order $O(\delta t)$ and smaller are neglected.

12. (i) What is the approximate value $P_{approx,r=0,q=0}$ of an at-the-money put option on a non–dividend–paying underlying asset with spot price $S = 60$, volatility $\sigma = 0.25$, and maturity $T = 1$ year, if the constant risk–free interest rate is $r = 0$?

(ii) Compute the Black–Scholes value $P_{BS,r=0,q=0}$ of the put option, and estimate the relative approximate error

$$\frac{|P_{BS,r=0,q=0} - P_{approx,r=0,q=0}|}{P_{BS,r=0,q=0}}.$$

(iii) Assume that $r = 0.06$ and $q = 0.03$. Compute the approximate value $P_{approx,r=0.06,q=0.03}$ of an ATM put option and estimate the relative approximate error

$$\frac{|P_{BS,r=0.06,q=0.03} - P_{approx,r=0.06,q=0.03}|}{P_{BS,r=0.06,q=0.03}}, \tag{6.11}$$

where $P_{BS,r=0.06,q=0.03}$ is the Black–Scholes value of the put option.

13. It is interesting to note that the approximate formulas

$$C \approx \sigma S \sqrt{\frac{T}{2\pi}} \left(1 - \frac{(r+q)T}{2}\right) + \frac{(r-q)T}{2} S;$$
$$P \approx \sigma S \sqrt{\frac{T}{2\pi}} \left(1 - \frac{(r+q)T}{2}\right) - \frac{(r-q)T}{2} S$$

for ATM call and put options do not satisfy the Put–Call parity:

$$P + Se^{-qT} - C = S\left(e^{-qT} - (r-q)T\right) \neq Se^{-rT} = Ke^{-rT}.$$

Based on the linear Taylor expansion $e^{-x} \approx 1 - x$, new approximation formulas for the price of ATM options which satisfy the Put–Call parity can be obtained by replacing rT and qT by $1 - e^{-rT}$ and $1 - e^{-qT}$, respectively. The resulting formulas are

$$C \approx \sigma S \sqrt{\frac{T}{2\pi}} \frac{e^{-qT} + e^{-rT}}{2} + \frac{S\left(e^{-qT} - e^{-rT}\right)}{2}; \tag{6.12}$$
$$P \approx \sigma S \sqrt{\frac{T}{2\pi}} \frac{e^{-qT} + e^{-rT}}{2} - \frac{S\left(e^{-qT} - e^{-rT}\right)}{2}. \tag{6.13}$$

(i) Show that the Put–Call parity is satisfied by the approximations (6.12) and (6.13).

(ii) Estimate how good the new approximation (6.13) is, for an ATM put with $S = 60$, $q = 0.03$, $\sigma = 0.25$, and $T = 1$, if $r = 0.06$, by computing the corresponding relative approximate error. Compare this error with the relative approximate error (6.11) found in a previous exercise.

14. Consider an ATM put option with strike 40 on a non–dividend paying asset with volatility 30%, and assume zero interest rates.

Compute the relative approximation error of the approximation

$$P \approx \sigma S \sqrt{\frac{T}{2\pi}}$$

if the put option expires in 3, 5, 10, and 20 years.

15. Consider an ATM put option with strike 40 on an asset with volatility 30% and paying 2% dividends continuously. Assume that the interest rates are constant at 4.5%. Compute the relative approximation error to the Black–Scholes value of the option of the approximate value

$$P_{approx,r\neq 0,q\neq 0} = \sigma S\sqrt{\frac{T}{2\pi}}\left(1 - \frac{(r+q)T}{2}\right) - \frac{(r-q)T}{2}S,$$

if the put option expires in 1, 3, 5, 10, and 20 years.

16. A five year bond worth 101 has duration 1.5 years and convexity equal to 2.5. Use both the formula

$$\frac{\Delta B}{B} \approx -D\Delta y,$$

which does not include any convexity adjustment, and the formula

$$\frac{\Delta B}{B} \approx -D\Delta y + \frac{1}{2}C(\Delta y)^2,$$

to find the price of the bond if the yield increases by ten basis points (i.e., 0.001), fifty basis points, one percent, and two percent, respectively.

17. Consider a bond portfolio worth \$50mil with DV01 equal to \$10,000 and dollar convexity equal to \$400mil.

(i) Assume that the yield curve moves up by fifty basis points. What is the new value of your bond portfolio?

(ii) The following bonds are available for trading:

	Principal	Value	Duration	Convexity
Bond 1	1000	1100	2.5	12
Bond 2	100	106	3	9

How do you immunize the portfolio?

18. (i) If the current zero rate curve is

$$r_1(0,t) = 0.025 + \frac{1}{100}\exp\left(-\frac{t}{100}\right) + \frac{t}{100(t+1)},$$

find the yield of a four year semiannual coupon bond with coupon rate 6%. Assume that interest is compounded continuously and that the face value of the bond is 100.

(ii) If the zero rates have a parallel shift up by 10, 20, 50, 100, and 200 basis points, respectively, i.e., if the zero rate curve changes from $r_1(0,t)$ to $r_2(0,t) = r_1(0,t) + dr$, with $dr = \{0.001, 0.002, 0.005, 0.01, 0.02\}$, find out by how much does the yield of the bond increase in each case.

Note: In general, a small parallel shift in the zero rate curves results in a shift of similar size and direction for the yield of most bonds (possibly with the exception of bonds with long maturity). This assumption will be tested for the bond considered here for parallel shifts ranging from small shifts (ten basis points) to large shifts (200 basis points).

6.2 Solutions to Chapter 6 Exercises

Problem 1: Show that the cubic Taylor approximation of $\sqrt{1+x}$ around 0 is

$$\sqrt{1+x} \approx 1 + \frac{x}{2} - \frac{x^2}{8} + \frac{x^3}{16}.$$

Solution: Recall that the cubic Taylor approximation of the function $f(x)$ around the point $a = 0$ is

$$\begin{aligned} f(x) &\approx f(a) + (x-a)f'(a) + \frac{(x-a)^2}{2}f''(a) + \frac{(x-a)^3}{6}f^{(3)}(a) \\ &= f(0) + xf'(0) + \frac{x^2}{2}f''(0) + \frac{x^3}{6}f^{(3)}(0). \end{aligned} \quad (6.14)$$

For $f(x) = \sqrt{1+x}$, we find that

$$f'(x) = \frac{1}{2\sqrt{1+x}}; \quad f''(x) = -\frac{1}{4(1+x)^{3/2}}; \quad f^{(3)}(x) = \frac{3}{8(1+x)^{5/2}},$$

and therefore

$$f(0) = 1; \; f'(0) = \frac{1}{2}; \; f''(0) = -\frac{1}{4}; \; f^{(3)}(0) = \frac{3}{8}. \quad (6.15)$$

From (6.14) and (6.15) we conclude that

$$\sqrt{1+x} \approx 1 + \frac{x}{2} - \frac{x^2}{8} + \frac{x^3}{16}. \quad \square$$

Problem 2: Use the Taylor series expansion of the function e^x to find the value of $e^{0.25}$ with six decimal digits accuracy.

Solution: Recall that the Taylor series expansion of the function $f(x) = e^x$ around 0 converges to e^x at all points $x \in \mathbb{R}$, i.e.,

$$e^x = \sum_{k=0}^{\infty} \frac{x^k}{k!}, \quad \forall\, x \in \mathbb{R}.$$

For $x = 0.25$ we find that

$$e^{0.25} = \sum_{k=0}^{\infty} \frac{(0.25)^k}{k!},$$

i.e.,

$$e^{0.25} = \lim_{n\to\infty} x_n, \quad \text{where} \quad x_n = \sum_{k=0}^{n} \frac{(0.25)^k}{k!}, \quad \forall\, n \geq 0.$$

Note that the sequence $\{x_n\}_{n=0:\infty}$ is increasing. It is then enough to compute x_0, x_1, x_2, ..., until the first seven decimal digits of these terms are the same, in order to find the first six decimal digits of $e^{0.25}$. We find that

$$\begin{array}{lll} x_0 = 1; & x_1 = 1.25; & x_2 = 1.28125; \\ x_3 = 1.28385417; & x_4 = 1.28401698; & x_5 = 1.28402507; \\ x_6 = 1.28402540; & x_7 = 1.28402541, & \end{array}$$

and conclude that

$$e^{0.25} \approx 1.284025. \quad \square$$

Problem 3: Show that

$$e^{-x} - \frac{1}{1+x} = O(x^2), \quad \text{as} \quad x \to 0.$$

Solution: The quadratic Taylor approximations of e^{-x} and $\frac{1}{1+x}$ are

$$\begin{aligned} e^{-x} &= 1 - x + \frac{x^2}{2} + O(x^3), \quad \text{as} \quad x \to 0; \\ \frac{1}{1+x} &= 1 - x + x^2 + O(x^3), \quad \text{as} \quad x \to 0. \end{aligned}$$

Therefore,

$$e^{-x} - \frac{1}{1+x} = -\frac{x^2}{2} + O(x^3) = O(x^2), \quad \text{as} \quad x \to 0.$$

Note that we implicitly proved that

$$e^{-x} - \frac{1}{1+x} = -\frac{x^2}{2} + O(x^3), \quad \text{as} \quad x \to 0. \quad \square$$

Problem 4: (i) Let $g(x)$ be an infinitely differentiable function. Find the linear and quadratic Taylor approximations of $e^{g(x)}$ around the point 0.

(ii) Use the result above to compute the quadratic Taylor approximation around 0 of $e^{(x+1)^2}$.

(iii) Compute the quadratic Taylor approximation around 0 of $e^{(x+1)^2}$ by using Taylor approximations of e^x and e^{x^2}.

Solution: (i) Let $f(x) = e^{g(x)}$. Then

$$f'(x) = g'(x)e^{g(x)} \quad \text{and} \quad f''(x) = (g''(x) + (g'(x))^2)e^{g(x)}.$$

The linear Taylor approximation

$$f(x) = f(0) + xf'(0) + O(x^2), \quad \text{as} \quad x \to 0,$$

can be written as

$$e^{g(x)} = e^{g(0)} + xe^{g(0)}g'(0) + O(x^2), \quad \text{as} \quad x \to 0.$$

The quadratic Taylor approximation

$$f(x) = f(0) + xf'(0) + \frac{x^2}{2}f''(0) + O(x^3), \quad \text{as} \quad x \to 0,$$

becomes

$$e^{g(x)} = e^{g(0)} + xe^{g(0)}g'(0) + x^2 e^{g(0)}\frac{g''(0) + (g'(0))^2}{2} + O(x^3), \tag{6.16}$$

as $x \to 0$.

(ii) Let $g(x) = (x+1)^2$. Then, $g'(x) = 2(x+1)$ and $g''(x) = 2$. Therefore, $g(0) = 1$, $g'(0) = 2$, $g''(0) = 2$, and, by substituting in (6.16), we find that

$$e^{(x+1)^2} = e + 2ex + 3ex^2 + O(x^3), \quad \text{as} \quad x \to 0.$$

(iii) Note that

$$e^{(x+1)^2} = e \cdot e^{2x} \cdot e^{x^2}. \tag{6.17}$$

Using the quadratic Taylor approximations

$$\begin{aligned} e^{2x} &= 1 + 2x + \frac{(2x)^2}{2} \\ &= 1 + 2x + 2x^2 + O(x^3), \quad \text{as} \quad x \to 0; \\ e^{x^2} &= 1 + x^2 + O(x^4), \quad \text{as} \quad x \to 0, \end{aligned}$$

it follows from (6.17) that

$$\begin{aligned} e^{(x+1)^2} &= e \cdot (1 + 2x + 2x^2 + O(x^3)) \cdot (1 + x^2 + O(x^4)) \\ &= e + 2ex + 3ex^2 + O(x^3), \quad \text{as} \quad x \to 0. \quad \square \end{aligned}$$

Problem 5: Find the Taylor series expansion of the functions

$$\ln(1 - x^2) \quad \text{and} \quad \frac{1}{1 - x^2}$$

around the point 0, using the Taylor series expansions of

$$\ln(1 - x) \quad \text{and} \quad \frac{1}{1 - x}$$

around 0.

Solution: Recall that

$$\ln(1-x) = -\sum_{k=1}^{\infty}\frac{x^k}{k} = -x - \frac{x^2}{2} - \frac{x^3}{3} - \frac{x^4}{4} - \dots, \ \forall\, x \in [-1,1);$$
$$\frac{1}{1-x} = \sum_{k=0}^{\infty} x^k = 1 + x + x^2 + x^3 + \dots, \ \forall\, x \in (-1,1).$$

By substituting x^2 for x in the Taylor expansions above, where $|x| < 1$, we find that

$$\ln(1-x^2) = -\sum_{k=1}^{\infty}\frac{x^{2k}}{k} = -x^2 - \frac{x^4}{2} - \frac{x^6}{3} - \frac{x^8}{4} - \dots, \ \forall\, x \in (-1,1);$$
$$\frac{1}{1-x^2} = \sum_{k=0}^{\infty} x^{2k} = 1 + x^2 + x^4 + x^6 + \dots, \ \forall\, x \in (-1,1). \quad \square$$

Problem 6: Prove that

$$\left(1+\frac{1}{x}\right)^{x+\frac{1}{2}-\frac{1}{12x}} < e < \left(1+\frac{1}{x}\right)^{x+\frac{1}{2}}, \quad \forall\, x \geq 1. \qquad (6.18)$$

Solution: Recall from (6.24) that

$$\ln\left(\frac{1+y}{1-y}\right) = 2y + \frac{2y^3}{3} + \frac{2y^5}{5} + \dots, \ \forall\, y \in (-1,1). \qquad (6.19)$$

For any $x \geq 1$, substitute $y = \frac{1}{2x+1}$ in (6.19) and obtain that

$$\ln\left(1+\frac{1}{x}\right) = \frac{2}{2x+1} + \frac{2}{3(2x+1)^3} + \frac{2}{5(2x+1)^5} + \dots, \ \forall\, x > 1,$$

which can also be written as

$$\left(x+\frac{1}{2}\right)\ln\left(1+\frac{1}{x}\right) = 1 + \frac{1}{3(2x+1)^2} + \frac{1}{5(2x+1)^4} + \dots, \ \forall\, x > 1. \quad (6.20)$$

From (6.20), we find that

$$1 < \left(x+\frac{1}{2}\right)\ln\left(1+\frac{1}{x}\right), \quad \forall\, x \geq 1,$$

which is equivalent to

$$e < \left(1+\frac{1}{x}\right)^{x+\frac{1}{2}}, \quad \forall\, x \geq 1. \tag{6.21}$$

The right inequality of (6.18) is therefore established.

From (6.20), we also find that

$$\begin{aligned}\left(x+\frac{1}{2}\right)\ln\left(1+\frac{1}{x}\right) &< 1+\frac{1}{3(2x+1)^2}\sum_{k=0}^{\infty}\frac{1}{(2x+1)^{2k}} \\ &= 1+\frac{1}{3(2x+1)^2}\cdot\frac{1}{1-\frac{1}{(2x+1)^2}} = 1+\frac{1}{12x(x+1)},\end{aligned}$$

and therefore that

$$\left(1+\frac{1}{x}\right)^{x+\frac{1}{2}} < e^{1+\frac{1}{12x(x+1)}}, \quad \forall\, x \geq 1. \tag{6.22}$$

Recall that

$$e < \left(1+\frac{1}{x}\right)^{x+1}, \quad \forall\, x \geq 1. \tag{6.23}$$

Using (6.23), we find from (6.22) that

$$\left(1+\frac{1}{x}\right)^{x+\frac{1}{2}} < e\cdot e^{\frac{1}{12x(x+1)}} < e\left(1+\frac{1}{x}\right)^{\frac{1}{12x}}, \quad \forall\, x \geq 1,$$

and we conclude that

$$\left(1+\frac{1}{x}\right)^{x+\frac{1}{2}-\frac{1}{12x}} < e, \quad \forall\, x \geq 1.$$

The left inequality of (6.18) is therefore established. □

Problem 7: Compute the Taylor series expansion of

$$\ln\left(\frac{1+x}{1-x}\right)$$

around the point 0, and find its radius of convergence.

Solution: Note that the function

$$\ln\left(\frac{1+x}{1-x}\right) = \ln(1+x) - \ln(1-x)$$

is not defined for $x = -1$ or $x = 1$. Therefore, the largest possible radius of convergence of its Taylor series expansion around 0 is 1.

The Taylor series expansions of the functions $\ln(1+x)$ and $\ln(1-x)$ are

$$\ln(1+x) = \sum_{k=1}^{\infty}(-1)^{k+1}\frac{x^k}{k} = x - \frac{x^2}{2} + \frac{x^3}{3} - \frac{x^4}{4} + \ldots, \ \forall\, x \in (-1, 1];$$

$$\ln(1-x) = -\sum_{k=1}^{\infty}\frac{x^k}{k} = -x - \frac{x^2}{2} - \frac{x^3}{3} - \frac{x^4}{4} - \ldots, \ \forall\, x \in [-1, 1),$$

and have radius of convergence equal to 1.

We conclude that the Taylor series expansion of $\ln\left(\frac{1+x}{1-x}\right)$ is

$$\begin{aligned}
\ln\left(\frac{1+x}{1-x}\right) &= \ln(1+x) - \ln(1-x) \\
&= \sum_{k=1}^{\infty}\left((-1)^{k+1}\frac{x^k}{k} + \frac{x^k}{k}\right) = \sum_{j=0}^{\infty}\frac{2x^{2j+1}}{2j+1} \\
&= 2x + \frac{2x^3}{3} + \frac{2x^5}{5} + \ldots, \ \forall\, x \in (-1, 1), \qquad (6.24)
\end{aligned}$$

and has radius of convergence equal to 1. □

Problem 8: (i) Find the radius of convergence of the series

$$1 + \frac{x^4}{2!} + \frac{x^8}{4!} + \frac{x^{12}}{6!} + \ldots \qquad (6.25)$$

(ii) Show that the series from (6.25) is the Taylor series expansion of the function

$$\frac{e^{x^2} + e^{-x^2}}{2}.$$

Solution: (i) The series (6.25) can be written as a power series as follows:

$$T(x) = \sum_{p=1}^{\infty} a_{4p}x^{4p}, \quad \text{with} \quad a_{4p} = \frac{1}{(2p)!}, \ \forall\, p \geq 1.$$

From Stirling's formula we know that

$$\lim_{k\to\infty}\frac{k!}{\left(\frac{k}{e}\right)^k\sqrt{2\pi k}} = 1.$$

It is then easy to see that

$$\lim_{k\to\infty} \frac{(k!)^{1/k}}{k} = \frac{1}{e},$$

and therefore that

$$\lim_{p\to\infty} \frac{2p}{((2p)!)^{1/2p}} = e \tag{6.26}$$

Using (6.26), we find that

$$\begin{aligned}\lim_{p\to\infty} |a_{4p}|^{1/4p} &= \lim_{p\to\infty} \frac{1}{((2p)!)^{1/4p}} = \lim_{p\to\infty} \left(\frac{2p}{((2p)!)^{1/2p}}\right)^{1/2} \cdot \frac{1}{(2p)^{1/2}} \\ &= \sqrt{e} \lim_{p\to\infty} \frac{1}{\sqrt{2p}} = 0.\end{aligned}$$

We conclude that the radius of convergence of the power series $T(x)$ is

$$R = \frac{1}{\limsup_{k\to\infty} |a_k|^{1/k}} = \frac{1}{\lim_{p\to\infty} |a_{4p}|^{1/4p}} = \infty,$$

which means that the series (6.25) is convergent for all $x \in \mathbb{R}$.
(ii) Using the Taylor series expansion

$$e^x = \sum_{k=0}^{\infty} \frac{x^k}{k!}, \quad \forall\, x \in \mathbb{R},$$

it is easy to see that

$$\begin{aligned}\frac{e^{x^2} + e^{-x^2}}{2} &= \frac{1}{2}\left(\sum_{k=0}^{\infty} \frac{(x^2)^k}{k!} + \sum_{k=0}^{\infty} \frac{(-x^2)^k}{k!}\right) \\ &= \frac{1}{2}\sum_{k=0}^{\infty} \frac{x^{2k} + (-1)^k x^{2k}}{k!} \\ &= \frac{1}{2}\sum_{j=0}^{\infty} \frac{x^{4j} + x^{4j}}{(2j)!} + \frac{1}{2}\sum_{j=0}^{\infty} \frac{x^{4j+2} + (-1)^{2j+1} x^{4j+2}}{(2j+1)!} \\ &= \sum_{j=0}^{\infty} \frac{x^{4j}}{(2j)!} + \frac{1}{2}\sum_{j=0}^{\infty} \frac{x^{4j+2} - x^{4j+2}}{(2j+1)!} \\ &= 1 + \frac{x^4}{2!} + \frac{x^8}{4!} + \frac{x^{12}}{6!} + \ldots,\end{aligned}$$

which is the same as the series (6.25). □

Problem 9: Let

$$T(x) = \sum_{k=1}^{\infty} \frac{(-1)^{k+1}\, x^k}{k}$$

be the Taylor series expansion of $f(x) = \ln(1+x)$. Our goal is to show that $T(x) = f(x)$ for all x such that $|x| < 1$.

Let

$$P_n(x) = \sum_{k=1}^{n} \frac{(-1)^{k+1}\, x^k}{k}$$

be the Taylor polynomial of degree n corresponding to $f(x)$. Since $T(x) = \lim_{n\to\infty} P_n(x)$, it follows that $f(x) = T(x)$ for all $|x| < 1$ if and only if

$$\lim_{n\to\infty} |f(x) - P_n(x)| = 0, \quad \forall\ |x| < 1. \tag{6.27}$$

(i) Show that, for any x,

$$f(x) - P_n(x) = \int_0^x \frac{(-1)^{n+2}\,(x-t)^n}{(1+t)^{n+1}}\, dt.$$

(ii) Show that, for any $0 \le x < 1$,

$$|f(x) - P_n(x)| \le x^n \ln(1+x)$$

and prove that (6.27) holds for all x such that $0 \le x < 1$.

(iii) Assume that $-1 < x < 0$. Show that

$$|f(x) - P_n(x)| \le (-x)^n |\ln(1+x)|$$

and conclude that (6.27) holds true for all x such that $-1 < x < 0$.

Solution: (i) From the integral formula for the Taylor approximation error we know that

$$f(x) - P_n(x) = \int_0^x \frac{(x-t)^n}{n!}\, f^{(n+1)}(t)\, dt. \tag{6.28}$$

Since $f(x) = \ln(1+x)$, we obtain by induction that the derivatives of $f(x)$ are

$$f^{(k)}(x) = \frac{(-1)^{k+1}(k-1)!}{(1+x)^k} \quad \forall\ k \ge 1. \tag{6.29}$$

From (6.28) and (6.29) it follows that

$$\begin{aligned} f(x) - P_n(x) &= \int_0^x \frac{(x-t)^n}{n!} \cdot \frac{(-1)^{n+2} n!}{(1+t)^{n+1}}\, dt \\ &= \int_0^x \frac{(-1)^{n+2}(x-t)^n}{(1+t)^{n+1}}\, dt. \end{aligned} \tag{6.30}$$

(ii) Let $x \in [0,1)$. By taking absolute values in (6.30) and using the fact that

$$\frac{x-t}{1+t} \le x, \quad \forall\, 0 \le t \le x < 1,$$

we obtain that

$$\begin{aligned} |f(x) - P_n(x)| &= \int_0^x \left(\frac{x-t}{1+t}\right)^n \frac{1}{1+t}\, dt \;\le\; x^n \int_0^x \frac{1}{1+t}\, dt \\ &= x^n \ln(1+x), \quad \forall\, 0 \le x < 1. \end{aligned}$$

We conclude that

$$\lim_{n\to\infty} |f(x) - P_n(x)| \;=\; 0, \quad \forall\, x \in [0,1). \tag{6.31}$$

(iii) Assume that $x \in (-1,0)$ and let $s = -x$. From (6.30), it follows that

$$\begin{aligned} f(x) - P_n(x) &= \int_0^x \frac{(-1)^{n+2}(x-t)^n}{(1+t)^{n+1}}\, dt \;=\; \int_0^{-s} \frac{(-1)^{n+2}(-s-t)^n}{(1+t)^{n+1}}\, dt \\ &= \int_0^{-s} \frac{(-1)^{2n+2}(s+t)^n}{(1+t)^{n+1}}\, dt \;=\; \int_0^{-s} \frac{(s+t)^n}{(1+t)^{n+1}}\, dt. \end{aligned}$$

Using the substitution $t = -z$, we obtain that

$$f(x) - P_n(x) \;=\; -\int_0^s \frac{(s-z)^n}{(1-z)^{n+1}}\, dz.$$

By taking absolute values and using the fact that

$$\frac{s-z}{1-z} \le s, \quad \forall\, 0 \le z \le s < 1,$$

we find that

$$\begin{aligned} |f(x) - P_n(x)| &= \int_0^s \frac{(s-z)^n}{(1-z)^{n+1}}\, dz \;\le\; \int_0^s \frac{s^n}{1-z}\, dz \\ &= s^n \left(-\ln(1-z)\right)\Big|_{z=0}^{z=s} \\ &= -s^n \ln(1-s) \;=\; s^n |\ln(1-s)| \\ &= (-x)^n |\ln(1+x)|; \end{aligned}$$

recall that, by definition $s = -x$.

Note that, for any $x \in (-1,0)$,

$$\lim_{n\to\infty} (-x)^n |\ln(1+x)| \;=\; 0.$$

We conclude that

$$\lim_{n\to\infty} |f(x) - P_n(x)| \;=\; 0, \quad \forall\, x \in (-1,0). \tag{6.32}$$

From (6.31) and (6.32) we obtain that

$$\lim_{n\to\infty} |f(x) - P_n(x)| = 0, \quad \forall\, x \in (-1,1),$$

and conclude that the Taylor series expansion of the function $f(x) = \ln(1+x)$ converges to $f(x)$ for any x with $|x| < 1$. □

Problem 10: The goal of this exercise is to compute

$$\int_0^1 \ln(1-x)\ln(x)\, dx. \tag{6.33}$$

(i) Show that

$$\lim_{x\searrow 0} (\ln(1-x)\ln(x)) = \lim_{x\nearrow 1} (\ln(1-x)\ln(x)) = 0,$$

and conclude that the integral (6.33) can be regarded as a definite integral.
(ii) Use the Taylor series expansion of $\ln(1-x)$ for $|x| < 1$ to show that

$$\int_0^1 \ln(1-x)\ln(x)\, dx = -\sum_{n=1}^{\infty} \frac{1}{n} \int_0^1 x^n \ln(x)\, dx.$$

(iii) Prove that

$$\int_0^1 \ln(1-x)\ln(x)\, dx = \sum_{k=1}^{\infty} \frac{1}{n(n+1)^2}.$$

(iv) Use that fact that

$$\sum_{k=1}^{\infty} \frac{1}{n^2} = \frac{\pi^2}{6}$$

to obtain that

$$\int_0^1 \ln(1-x)\ln(x)\, dx = 2 - \frac{\pi^2}{6}.$$

Solution: (i) First of all, note that

$$\lim_{x\searrow 0} (\ln(1-x)\ln(x)) = \lim_{x\nearrow 1} (\ln(1-x)\ln(x)). \tag{6.34}$$

We compute the left hand side limit of (6.34) by changing it to a limit to infinity corresponding to $y = \frac{1}{x}$ as follows:

$$\lim_{x\searrow 0} (\ln(1-x)\ln(x)) = \lim_{y\to\infty} \ln\left(1 - \frac{1}{y}\right) \ln\left(\frac{1}{y}\right)$$

$$
\begin{aligned}
&= \lim_{y\to\infty} \left(\ln\left(\left(1-\frac{1}{y}\right)^y \right) \cdot \frac{1}{y} \right) (-\ln(y)) \\
&= -\lim_{y\to\infty} \ln\left(\left(1-\frac{1}{y}\right)^y \right) \cdot \frac{\ln(y)}{y}. \qquad (6.35)
\end{aligned}
$$

Recall that

$$\lim_{y\to\infty} \left(1-\frac{1}{y}\right) = \frac{1}{e}$$

and therefore

$$\lim_{y\to\infty} \ln\left(\left(1-\frac{1}{y}\right)^y \right) = -1. \qquad (6.36)$$

From (6.35) and (6.36) it follows that

$$\lim_{x\searrow 0} (\ln(1-x)\ln(x)) = \lim_{y\to\infty} \frac{\ln(y)}{y} = 0.$$

We conclude that

$$\lim_{x\searrow 0} (\ln(1-x)\ln(x)) = \lim_{x\nearrow 1} (\ln(1-x)\ln(x)) = 0.$$

The integral (6.33) is equal to the definite integral between 0 and 1 of the continuous function $g : [0,1] \to \mathbb{R}$ given by

$$g(0) = g(1) = 0; \quad g(x) = \ln(1-x)\ln(x), \ \ \forall\, 0 < x < 1.$$

(ii) The Taylor series expansion of $\ln(1-x)$, i.e.,

$$\ln(1-x) = -\sum_{k=1}^{\infty} \frac{x^k}{k} \ \ \forall\, x \in (-1,1),$$

is absolutely convergent to $\ln(1-x)$. Then,

$$
\begin{aligned}
\int_0^1 \ln(1-x)\ln(x)\,dx &= -\int_0^1 \left(\sum_{n=1}^{\infty} \frac{x^n \ln(x)}{n} \right) dx \\
&= -\sum_{n=1}^{\infty} \frac{1}{n} \int_0^1 x^n \ln(x)\,dx. \qquad (6.37)
\end{aligned}
$$

(iii) Using integration by parts, it is easy to see that

$$\int x^n \ln(x)\,dx = \frac{x^{n+1}\ln(x)}{n+1} - \frac{x^{n+1}}{(n+1)^2} + C.$$

Then,

$$\begin{aligned}\int_0^1 x^n \ln(x)\, dx &= \left(\frac{x^{n+1}\ln(x)}{n+1} - \frac{x^{n+1}}{(n+1)^2}\right)\Bigg|_0^1 \\ &= -\frac{1}{(n+1)^2} - \frac{1}{n+1}\lim_{x\searrow 0}\left(x^{n+1}\ln(x)\right) \\ &= -\frac{1}{(n+1)^2} \quad \forall\, n \geq 1. \end{aligned} \tag{6.38}$$

From (6.37) and (6.38), it follows that

$$\int_0^1 \ln(1-x)\ln(x)\, dx = \sum_{n=1}^{\infty} \frac{1}{n(n+1)^2}. \tag{6.39}$$

(iv) Note that

$$\frac{1}{n(n+1)} = \frac{1}{n} - \frac{1}{n+1}.$$

Then, it is easy to see that

$$\begin{aligned}\frac{1}{n(n+1)^2} &= \frac{1}{n(n+1)} \cdot \frac{1}{n+1} = \frac{1}{n(n+1)} - \frac{1}{(n+1)^2} \\ &= \frac{1}{n} - \frac{1}{n+1} - \frac{1}{(n+1)^2}.\end{aligned}$$

Therefore,

$$\begin{aligned}\sum_{n=1}^{\infty} \frac{1}{n(n+1)^2} &= \sum_{n=1}^{\infty}\left(\frac{1}{n} - \frac{1}{n+1}\right) - \sum_{n=2}^{\infty}\frac{1}{n^2} \\ &= 1 - \left(\frac{\pi^2}{6} - 1\right) = 2 - \frac{\pi^2}{6}.\end{aligned} \tag{6.40}$$

Here, we used the fact that

$$\sum_{k=1}^{\infty} \frac{1}{n^2} = \frac{\pi^2}{6}$$

and the telescoping series

$$\sum_{n=1}^{\infty}\left(\frac{1}{n} - \frac{1}{n+1}\right) = \lim_{N\to\infty}\sum_{n=1}^{N}\left(\frac{1}{n} - \frac{1}{n+1}\right) = \lim_{N\to\infty} 1 - \frac{1}{N+1} = 1.$$

From (6.39) and (6.40), we conclude that

$$\int_0^1 \ln(1-x)\ln(x)\, dx = 2 - \frac{\pi^2}{6}. \quad \square$$

Problem 11: In the Cox-Ross-Rubinstein parametrization for a binomial tree, the up and down factors u and d, and the risk–neutral probability p of the price going up during one time step are

$$u = A + \sqrt{A^2 - 1}; \tag{6.41}$$

$$d = A - \sqrt{A^2 - 1}; \tag{6.42}$$

$$p = \frac{e^{r\delta t} - d}{u - d}, \tag{6.43}$$

where

$$A = \frac{1}{2}\left(e^{-r\delta t} + e^{(r+\sigma^2)\delta t}\right).$$

Use Taylor expansions to show that, for a small time step δt, u, d and p may be approximated by

$$u = e^{\sigma\sqrt{\delta t}}; \tag{6.44}$$

$$d = e^{-\sigma\sqrt{\delta t}}; \tag{6.45}$$

$$p = \frac{1}{2} + \frac{1}{2}\left(\frac{r}{\sigma} - \frac{\sigma}{2}\right)\sqrt{\delta t}. \tag{6.46}$$

In other words, write the Taylor expansions for (6.41–6.43) and for (6.44–6.46) and show that they are identical if all the terms of order $O(\delta t)$ and smaller are neglected.

Solution: We will show that the Taylor expansions for (6.41–6.43) and for (6.44–6.46) are identical if all the terms of order $O((\delta t)^{3/2})$ and smaller are neglected. In other words, we will show that

$$u = e^{\sigma\sqrt{\delta t}} + O((\delta t)^{3/2}); \tag{6.47}$$

$$d = e^{-\sigma\sqrt{\delta t}} + O((\delta t)^{3/2}). \tag{6.48}$$

$$p = \frac{1}{2} + \frac{1}{2\sigma}(r - \frac{\sigma^2}{2})\sqrt{\delta t} \ + \ O((\delta t)^{3/2}), \tag{6.49}$$

Recall the following Taylor approximations:

$$e^x = 1 + x + \frac{x^2}{2} + O(x^3), \quad \text{as } x \to 0;$$

$$\sqrt{1+x} = 1 + \frac{x}{2} + O(x^2), \quad \text{as } x \to 0;$$

$$\sqrt{1-x} = 1 - \frac{x}{2} + O(x^2), \quad \text{as } x \to 0.$$

In particular, note that

$$\sqrt{1 + O(\delta t)} = 1 + \frac{O(\delta t)}{2} + O((\delta t)^2).$$

Then,

$$\begin{aligned}
A &= \frac{1}{2}\left(e^{-r\delta t} + e^{(r+\sigma^2)\delta t}\right) \\
&= \frac{1}{2}\left(1 - r\delta t + O((\delta t)^2) \ + \ 1 + (r+\sigma^2)\delta t + O((\delta t)^2)\right) \\
&= 1 + \frac{\sigma^2 \delta t}{2} + O((\delta t)^2); \\
A^2 - 1 &= \left(1 + \frac{\sigma^2 \delta t}{2} + O((\delta t)^2)\right)^2 - 1 \\
&= \sigma^2 \delta t + O((\delta t)^2; \\
\sqrt{A^2 - 1} &= \sqrt{\sigma^2 \delta t + O((\delta t)^2)} \ = \ \sigma\sqrt{\delta t}\ \sqrt{1 + O(\delta t)} \\
&= \sigma\sqrt{\delta t}\cdot\left(1 + \frac{O(\delta t)}{2} \ + \ O((\delta t)^2)\right) \\
&= \sigma\sqrt{\delta t} + O((\delta t)^{3/2})
\end{aligned}$$

and

$$\begin{aligned}
u &= A + \sqrt{A^2 - 1} \\
&= 1 + \frac{\sigma^2 \delta t}{2} + O((\delta t)^2) \ + \ \sigma\sqrt{\delta t} + O((\delta t)^{3/2}) \\
&= 1 + \sigma\sqrt{\delta t} + \frac{\sigma^2 \delta t}{2} \ + \ O((\delta t)^{3/2}); \\
d &= A - \sqrt{A^2 - 1} \\
&= 1 - \sigma\sqrt{\delta t} + \frac{\sigma^2 \delta t}{2} \ + \ O((\delta t)^{3/2}).
\end{aligned}$$

Since

$$\begin{aligned}
e^{\sigma\sqrt{\delta t}} &= 1 + \sigma\sqrt{\delta t} + \frac{(\sigma\sqrt{\delta t})^2}{2} \ + \ O((\sigma\sqrt{\delta t})^3) \\
&= 1 + \sigma\sqrt{\delta t} + \frac{\sigma^2 \delta t}{2} \ + \ O((\delta t)^{3/2}); \\
e^{-\sigma\sqrt{\delta t}} &= 1 - \sigma\sqrt{\delta t} + \frac{\sigma^2 \delta t}{2} \ + \ O((\delta t)^{3/2}),
\end{aligned}$$

we conclude that

$$\begin{aligned}
u &= e^{\sigma\sqrt{\delta t}} + O((\delta t)^{3/2}); \\
d &= e^{-\sigma\sqrt{\delta t}} + O((\delta t)^{3/2}).
\end{aligned}$$

Therefore, (6.47) and (6.48) are established.

Finally,

$$\begin{aligned} p &= \frac{e^{r\delta t} - d}{u - d} \\ &= \frac{\left(1 + r\delta t + O((r\delta t)^2)\right) - \left(1 - \sigma\sqrt{\delta t} + \frac{\sigma^2\delta t}{2} + O((\delta t)^{3/2})\right)}{\left(1 + \sigma\sqrt{\delta t} + \frac{\sigma^2\delta t}{2} + O((\delta t)^{3/2})\right) - \left(1 - \sigma\sqrt{\delta t} + \frac{\sigma^2\delta t}{2} + O((\delta t)^{3/2})\right)} \\ &= \frac{\sigma\sqrt{\delta t} + (r - \frac{\sigma^2}{2})\delta t + O((\delta t)^{3/2})}{2\sigma\sqrt{\delta t} + O((\delta t)^{3/2})} = \frac{\sigma + (r - \frac{\sigma^2}{2})\sqrt{\delta t} + O(\delta t)}{2\sigma + O(\delta t)} \\ &= \frac{1}{2} + \frac{1}{2\sigma}(r - \frac{\sigma^2}{2})\sqrt{\delta t} + O((\delta t)^{3/2}), \end{aligned}$$

which is what we wanted to show; cf. (6.49) . □

Problem 12: (i) What is the approximate value $P_{approx,r=0,q=0}$ of an at-the-money put option on a non–dividend–paying underlying asset with spot price $S = 60$, volatility $\sigma = 0.25$, and maturity $T = 1$ year, if the constant risk–free interest rate is $r = 0$?

(ii) Compute the Black–Scholes value $P_{BS,r=0,q=0}$ of the put option, and estimate the relative approximate error

$$\frac{|P_{BS,r=0,q=0} - P_{approx,r=0,q=0}|}{P_{BS,r=0,q=0}}.$$

(iii) Assume that $r = 0.06$ and $q = 0.03$. Compute the approximate value $P_{approx,r=0.06,q=0.03}$ of an ATM put option and estimate the relative approximate error

$$\frac{|P_{BS,r=0.06,q=0.03} - P_{approx,r=0.06,q=0.03}|}{P_{BS,r=0.06,q=0.03}}, \tag{6.50}$$

where $P_{BS,r=0.06,q=0.03}$ is the Black–Scholes value of the put option.

Solution: (i) Using the approximate formula

$$P_{approx,r=0,q=0} = \sigma S\sqrt{\frac{T}{2\pi}},$$

we obtain that

$$P_{approx,r=0,q=0} = 5.984134.$$

(ii) From the Black–Scholes formula, we find that

$$P_{BS,r=0,q=0} = 5.968592,$$

and therefore

$$\frac{|P_{BS,r=0,q=0} - P_{approx,r=0,q=0}|}{P_{BS,r=0,q=0}} = 0.002604 = 0.26\%.$$

(iii) Using the approximate formula

$$P_{approx,r\neq 0,q\neq 0} = \sigma S\sqrt{\frac{T}{2\pi}}\left(1 - \frac{(r+q)T}{2}\right) - \frac{(r-q)T}{2}S,$$

we obtain that

$$P_{approx,r=0.06,q=0.03} = 4.814848.$$

From the Black–Scholes formula, we find that

$$P_{BS,r=0.06,q=0.03} = 4.886985,$$

and therefore

$$\frac{|P_{BS,r=0.06,q=0.03} - P_{approx,r=0.06,q=0.03}|}{P_{BS,r=0.06,q=0.03}} = 0.014761 = 1.4761\%. \quad \square$$

Problem 13: It is interesting to note that the approximate formulas

$$C \approx \sigma S\sqrt{\frac{T}{2\pi}}\left(1 - \frac{(r+q)T}{2}\right) + \frac{(r-q)T}{2}S;$$

$$P \approx \sigma S\sqrt{\frac{T}{2\pi}}\left(1 - \frac{(r+q)T}{2}\right) - \frac{(r-q)T}{2}S$$

for ATM call and put options do not satisfy the Put–Call parity:

$$P + Se^{-qT} - C = S\left(e^{-qT} - (r-q)T\right) \neq Se^{-rT} = Ke^{-rT}.$$

Based on the linear Taylor expansion $e^{-x} \approx 1 - x$, new approximation formulas for the price of ATM options which satisfy the Put–Call parity can be obtained by replacing rT and qT by $1 - e^{-rT}$ and $1 - e^{-qT}$, respectively. The resulting formulas are

$$C \approx \sigma S\sqrt{\frac{T}{2\pi}}\,\frac{e^{-qT} + e^{-rT}}{2} + \frac{S\left(e^{-qT} - e^{-rT}\right)}{2}; \tag{6.51}$$

$$P \approx \sigma S\sqrt{\frac{T}{2\pi}}\,\frac{e^{-qT} + e^{-rT}}{2} - \frac{S\left(e^{-qT} - e^{-rT}\right)}{2}. \tag{6.52}$$

(i) Show that the Put–Call parity is satisfied by the approximations (6.51) and (6.52).

(ii) Estimate how good the new approximation (6.52) is, for an ATM put with $S = 60$, $q = 0.03$, $\sigma = 0.25$, and $T = 1$, if $r = 0.06$, by computing the corresponding relative approximate error. Compare this error with the relative approximate error (6.50) found in a previous exercise.

Solution: (i) From (6.51) and (6.52), it is easy to see that

$$P - C + Se^{-qT} = -2\,\frac{S\left(e^{-qT} - e^{-rT}\right)}{2} + Se^{-qT} = Se^{-rT} = Ke^{-rT},$$

since $K = S$ for ATM options.

(ii) Using the new approximation formula (6.52), we obtain that

$$P_{approx\_new,r=0.06,q=0.03} = 4.861031.$$

The Black–Scholes value of the put option is $P_{BS,r=0.06,q=0.03} = 4.886985$, and therefore

$$\frac{\left|P_{BS,r=0.06,q=0.03} - P_{approx\_new,r=0.06,q=0.03}\right|}{P_{BS,r=0.06,q=0.03}} = 0.005311 = 0.5311\%.$$

Recall from Problem 12 that the approximation error corresponding to the original approximation formula is 1.4761%. For this particular example, the new approximation formula is more accurate. □

Problem 14: Consider an ATM put option with strike 40 on a non–dividend paying asset with volatility 30%, and assume zero interest rates.

Compute the relative approximation error of the approximation

$$P \approx \sigma S\sqrt{\frac{T}{2\pi}}$$

if the put option expires in 1, 3, 5, 10, and 20 years.

Solution: We expect the precision of the approximation formula for ATM options to decrease as the maturity of the option increases. This is, indeed, the case:

T	P_{approx}	P_{BS}	Approximation Error
1	4.787307	4.769417	0.38%
3	8.291860	8.199509	1.13%
5	10.704745	10.507368	1.88%
10	15.138795	14.589748	3.76%
20	21.409489	19.906608	7.55%

Here, $P_{approx} = \sigma S\sqrt{\frac{T}{2\pi}}$, and the Approximation Error is the relative approximation error given by

$$\frac{|P_{BS} - P_{approx}|}{P_{BS,r=0,q=0}}. \quad \square$$

Problem 15: Consider an ATM put option with strike 40 on an asset with volatility 30% and paying 2% dividends continuously. Assume that the interest rates are constant at 4.5%. Compute the relative approximation error of the approximation

$$P_{approx,r\neq 0,q\neq 0} = \sigma S\sqrt{\frac{T}{2\pi}}\left(1 - \frac{(r+q)T}{2}\right) - \frac{(r-q)T}{2}S$$

if the put option expires in 1, 3, 5, 10, and 20 years.

Solution: The approximate option values and the corresponding approximation errors are given below:

T	P_{approx}	P_{BS}	Error
1	4.1317	4.1491	0.42%
3	5.9834	6.1577	2.83%
5	6.4652	6.9714	7.26%
10	5.2187	7.3398	28.90%
20	-2.5067	6.0595	N/A

While the approximation formula is still within 3% of the Black–Scholes value when the maturity is three years or less, it deteriorates for long dated options, and even produces a negative value for the 20–years option. $\square$

Problem 16: A five year bond worth 101 has duration 1.5 years and convexity equal to 2.5. Use both the formula

$$\frac{\Delta B}{B} \approx -D\Delta y, \tag{6.53}$$

which does not include any convexity adjustment, and the formula

$$\frac{\Delta B}{B} \approx -D\Delta y + \frac{1}{2}C(\Delta y)^2, \tag{6.54}$$

to find the price of the bond if the yield increases by ten basis points (i.e., 0.001), fifty basis points, one percent, and two percent, respectively.

Solution: Denote by $B_{new,D}$ the approximate value given by formula (6.53) for the value of the bond corresponding to the new yield. Then, $\Delta B = B_{new,D} - B$ and, from (6.53), it follows that

$$B_{new,D} \;=\; B\,(1 - D\Delta y). \tag{6.55}$$

Similarly, let $B_{new,D,C}$ the approximate value for the value of the bond given by formula (6.54). We obtain that

$$B_{new,D,C} \;=\; B\left(1 - D\Delta y + \frac{C}{2}(\Delta y)^2\right). \tag{6.56}$$

Note that $B = 101$, $D = 1.5$, and $C = 2.5$ in (6.55) and (6.56). The approximate values are obtained for $\Delta y \in \{0.001, 0.005, 0.01, 0.02\}$ can be found in the table below, where the last column of the table represents the percent difference between the approximate value using duration alone, and the approximate value using both duration and convexity, i.e.,

$$\frac{B_{new,D,C} - B_{new,D}}{B_{new,D}}:$$

Δy	$B_{new,D}$	$B_{new,D,C}$	
0.0010	100.8485	100.8486	0.0001%
0.0050	100.2425	100.2457	0.0031%
0.01	99.4850	99.4976	0.0127%
0.02	97.9700	98.0205	0.0515%

Problem 17: Consider a bond portfolio worth \$50mil with DV01 equal to \$10,000 and dollar convexity equal to \$400mil.

(i) Assume that the yield curve moves up by fifty basis points. What is the new value of your bond portfolio?

(ii) The following bonds are available for trading:

	Principal	Value	Duration	Convexity
Bond 1	1000	1100	2.5	12
Bond 2	100	106	3	9

How do you immunize the portfolio?

Solution: Quick computation: for a one basis point shift up of the yield curve, the value of the bond portfolio decreases by an amount equal to its

DV01, i.e., \$10,000. Thus, a fifty basis points upward move of the yield curve will result in a loss of

$$50 \cdot \$10,000 \;=\; \$500,000$$

in the bond portfolio.

The new value of the portfolio will be approximately \$49.5mil.

A more precise approach includes a convexity adjustment. Recall that, if the yield curve experiences a small parallel shift of size δr, the change ΔV in the value of the bond portfolio can be approximated as

$$\Delta V \;\approx\; -D_{\$}(V)\delta r \;+\; \frac{C_{\$}(V)}{2}(\delta r)^2, \tag{6.57}$$

where $D_{\$}(V)$ and $C_{\$}(V)$ are the dollar duration and the dollar convexity of the portfolio, respectively.

The DV01 of a bond portfolio is equal to the dollar duration of the bond portfolio divided by 10,000, i.e.,

$$\mathrm{DV01}(V) \;=\; \frac{D_{\$}(V)}{10,000}.$$

Thus,

$$D_{\$}(V) \;=\; \mathrm{DV01}(V) \cdot 10,000 \;=\; \$100\text{mil}.$$

Using (6.57) with $\delta r = 50\text{bps} = 0.005$, $D_{\$}(V) = \$100\text{mil}$, and $C_{\$}(V) = \$400\text{mil}$, we obtain that the change in the value of the bond portfolio is

$$\begin{aligned} \Delta V &\approx -\$100\text{mil} \cdot 0.005 \;+\; \frac{\$400\text{mil}}{2} \cdot (0.005)^2 \\ &= -\$500,000 + \$5,000 \\ &= -\$495,000. \end{aligned}$$

The new value of the bond portfolio is

$$V_{new} \;=\; V_{old} + \Delta V \;\approx\; \$50\text{mil} - \$495,000 \;=\; \$49,505,000.$$

Note that the convexity adjustment to the change in the value of the portfolio, i.e., \$5,000, is much smaller than the change in the value of the portfolio given by DV01, i.e., \$500,000.

(ii) Denote by B_1 and B_2 the sizes of the positions taken in the two bonds available for trading. The value of the new portfolio is

$$\Pi \;=\; V \;+\; B_1 \;+\; B_2.$$

The dollar duration and the dollar convexity of the bond positions are

$$\begin{aligned} D_{\$}(B_1) &= B_1 D_1 = 2.5B_1; \quad & C_{\$}(B_1) &= B_1 C_1 = 12B_1; \\ D_{\$}(B_2) &= B_2 D_2 = 3B_2; \quad & C_{\$}(B_2) &= B_2 C_2 = 9B_2. \end{aligned}$$

Recall that $D_\$(V) = \$100\text{mil}$ and $C_\$(V) = \$400\text{mil}$. The dollar duration and the dollar convexity of the new portfolio are

$$\begin{aligned} D_\$(\Pi) &= D_\$(V) + D_\$(B_1) + D_\$(B_2) = \$100\text{mil} + 2.5B_1 + 3B_2; \\ C_\$(\Pi) &= C_\$(V) + C_\$(B_1) + C_\$(B_2) = \$400\text{mil} + 12B_1 + 9B_2. \end{aligned}$$

In order for $D_\$(\Pi)$ and for $C_\$(\Pi)$ to be equal to zero, the positions B_1 and B_2 must satisfy the following linear system:

$$\begin{cases} 2.5B_1 + 3B_2 = -\$100\text{mil}; \\ 12B_1 + 9B_2 = -\$400\text{mil}. \end{cases}$$

The solution of this system is

$$\begin{aligned} B_1 &= -\frac{\$200\text{mil}}{9} = -\$22,222,222; \\ B_2 &= -\frac{\$400\text{mil}}{27} = -\$14,814,815. \end{aligned}$$

These are the sizes of the bond positions we must add to the original portfolio in order to immunize it from small changes in the yield curve.

We conclude that, in order to immunize the given portfolio, we must short $\frac{22,222,222}{1100} \approx 20,202$ units of Bond 1, and short $\frac{14,814,815}{106} \approx 139,762$ units of Bond 2. □

Problem 18: (i) If the current zero rate curve is

$$r_1(0,t) = 0.025 + \frac{1}{100}\exp\left(-\frac{t}{100}\right) + \frac{t}{100(t+1)},$$

find the yield of a four year semiannual coupon bond with coupon rate 6%. Assume that interest is compounded continuously and that the face value of the bond is 100.

(ii) If the zero rates have a parallel shift up by 10, 20, 50, 100, and 200 basis points, respectively, i.e., if the zero rate curve changes from $r_1(0,t)$ to $r_2(0,t) = r_1(0,t) + dr$, with $dr = \{0.001, 0.002, 0.005, 0.01, 0.02\}$, find out by how much does the yield of the bond increase in each case.

Solution: (i) The bond provides coupon payments equal to 3 every six months until 3.5 years from now, and a final cash flow of 103 in four years. By discounting this cash flows to the present using the zero rate curve $r_1(0,t)$, we find that the value of the bond is

$$\begin{aligned} B_1 &= \sum_{i=1}^{7} 3\exp\left(-r_1\left(0,\frac{i}{2}\right)\frac{i}{2}\right) + 103\exp(-4r_1(0,4)) \\ &= 106.1995. \end{aligned} \tag{6.58}$$

The yield of the bond is found by solving the formula for the price of the bond in terms of its yield, i.e, by solving

$$B_1 = \sum_{i=1}^{7} 3 \exp\left(-y\,\frac{i}{2}\right) + 103 \exp(-4y) \tag{6.59}$$

for y, where B_1 is given by (6.58), i.e., $B_1 = 106.1995$. Using Newton's method, we obtain that the yield of the bond is

$$y = 0.042511 = 4.2511\%.$$

(ii) If the zero rates increase, the value of the bond decreases, and therefore the yield of the bond will increase. Our goal here is to investigate whether a parallel shift of the zero curve up by dr results in an increase of the yield of the bond also equal to dr.

When the zero rates increase from $r_1(0,t)$ to $r_2(0,t) = r_1(0,t) + dr$, the value of the bond decreases from B_1 given by (6.58) to

$$B_2 = \sum_{i=1}^{7} 3 \exp\left(-r_2\left(0,\frac{i}{2}\right)\frac{i}{2}\right) + 103 \exp(-4r_2(0,4)).$$

The new yield of the bond, denoted by y_2, will be larger than the initial yield y, and is obtained by solving

$$B_2 = \sum_{i=1}^{7} 3 \exp\left(-y_2\frac{i}{2}\right) + 103 \exp(-4y_2) \tag{6.60}$$

for y_2, where B_2 is given by (6.60).

For parallel shifts equal to

$$dr = \{0.001, 0.002, 0.005, 0.01, 0.02\},$$

we obtain the following bond prices and yields:

Zero rate shift dr	New bond price B_2	New yield y_2	Yield increase $y_2 - y$
10bps = 0.001	105.8150	0.043511	0.00099979
20bps = 0.002	105.4319	0.044510	0.00199957
50bps = 0.005	104.2915	0.047510	0.00499893
100bps = 0.01	102.4199	0.052509	0.00999784
200bps = 0.02	98.7829	0.062506	0.01999562

As expected, the increase of the yield of the bond is slightly smaller, but very close to, the parallel shift of the zero rate curve, i.e., $y_2 - y \approx dr$. □

Chapter 7

Finite Differences. Black–Scholes PDE.

7.1 Exercises

1. Let $f(x) = x^3e^x - 6e^x$. Show that the central finite difference approximation for $f'(0)$ is a fourth order approximation.

2. Let $N(x)$ be the cumulative density of the standard normal variable. Show that the forward finite difference approximation of $N'(0)$ is a second order approximation.

3. Find a second order finite difference approximation for $f'(a)$ using $f(a)$, $f(a+h)$, and $f(a+2h)$.

   Note: This type of approximation is needed, e.g., when discretizing a PDE with boundary conditions involving derivatives of the solution (also called Robin boundary conditions). For example, for continuously compounded average rate Asian call options, this type of finite difference approximation is used to discretize the boundary condition $\frac{\partial H}{\partial t} + \frac{\partial H}{\partial R} = 0$, for $R = 0$.

4. Let $f(x)$ be an infinitely many times differentiable function. Find a central finite difference approximation for the fourth derivative of f at a, i.e., for $f^{(4)}(a)$, using $f(a-2h)$, $f(a-h)$, $f(a)$, $f(a+h)$, and $f(a+2h)$. What is the order of this finite difference approximation?

5. The goal of this exercise is to emphasize the importance of symmetry in finite difference approximations. Recall that the central difference approximations for the first and second order derivatives are

$$f'(a) = \frac{f(a+h) - f(a-h)}{2h} + O\left(h^2\right);$$
$$f''(a) = \frac{f(a+h) - 2f(a) + f(a-h)}{h^2} + O\left(h^2\right),$$

as $h \to 0$. In other words, $f'(a)$ and $f''(a)$ are approximated to second order accuracy by using the value of f at the point a and at the points $a - h$ and $a + h$ that are symmetric with respect to a.

We investigate what happens if symmetry is not required.

(i) Find a second order finite difference approximation of $f'(a)$ using $f(a)$, $f(a - h)$ and $f(a + 2h)$.

(ii) Find a first order finite difference approximation of $f''(a)$ using $f(a)$, $f(a-h)$ and $f(a+2h)$. Note that, in general, a second order finite difference approximation of $f''(a)$ using $f(a)$, $f(a - h)$ and $f(a + 2h)$ does not exist.

Let $\beta < a < \gamma$ such that $a - \beta = C(\gamma - a)$, where C is a constant.

(iii) Find a finite difference approximation of $f'(a)$ using $f(a)$, $f(\beta)$, and $f(\gamma)$, which is second order in terms of $|\gamma - a|$, i.e., where the residual term is $O\left(|\gamma - a|^2\right)$.

(iv) Find a finite difference approximation of $f''(a)$ using $f(a)$, $f(\beta)$, and $f(\gamma)$, which is first order in terms of $|\gamma - a|$. Show that, in general, a second order finite difference approximation of $f''(a)$ using $f(a)$, $f(\beta)$ and $f(\gamma)$ is not possible, unless $a = \frac{\beta+\gamma}{2}$, i.e., unless β and γ are symmetric with respect to a.

6. A butterfly spread is made of a long position in a call option with strike $K - x$, a long position in a call option with strike $K + x$, and a short position in two calls with strike K. The options are on the same underlying asset and have the same maturities.

(i) Show that the value of the butterfly spread is

$$C(K + x) - 2C(K) + C(K - x),$$

where, e.g., $C(K + x)$ denotes the price of the call with strike $K + x$.

(ii) Show that, in the limiting case when x goes to 0, the **value** of a position in $\frac{1}{x^2}$ butterfly spreads as above converges to the second order partial derivative of the value of the option, C, with respect to strike K, i.e., show that

$$\lim_{x \searrow 0} \frac{C(K + x) - 2C(K) + C(K - x)}{x^2} = \frac{\partial^2 C}{\partial K^2}(K).$$

(iii) Show that, in the limiting case when $x \to 0$, the **payoff at maturity** of a position in $\frac{1}{x}$ butterfly spreads as above is going to approximate the payoff of a derivative security that pays 1 if the underlying asset expires at K, and 0 otherwise.

Note: A security that pays 1 in a certain state and 0 is any other state is called an Arrow-Debreu security, and its price is called the Arrow-Debreu price of that state. A position in $\frac{1}{x}$ butterfly spreads as above, with x small, is a synthetic way to construct an Arrow-Debreu security for the state $S(T) = K$.

7. A bull spread is made of a long position in a call option with strike K and a short position in a call option with strike $K + x$, both options being on the same underlying asset and having the same maturities. Let $C(K)$ and $C(K + x)$ be the values (at time t) of the call options with strikes K and $K + x$, respectively.

   (i) The **value** of a position in $\frac{1}{x}$ bull spreads is $\frac{C(K)-C(K+x)}{x}$. In the limiting case when x goes to 0, show that

$$\lim_{x \searrow 0} \frac{C(K) - C(K + x)}{x} = -\frac{\partial C}{\partial K}(K).$$

   (ii) Show that, in the limiting case when $x \to 0$, the **payoff at maturity** of a position in $\frac{1}{x}$ bull spreads as above is going to approximate the payoff of a derivative security that pays 1 if the price of the underlying asset at expiry is above K, and pays 0 if the price of the underlying asset at expiry is below K.

   Note: A position in $\frac{1}{x}$ bull spreads as above, with x small, is a synthetic way to construct a cash–or–nothing call maturing at time T.

8. Let

$$f(S) = \frac{C(K + x, T) - 2C(K, T) + C(K - x, T)}{x^2},$$

   where, e.g., $C(K, T)$ denotes the value of a plain vanilla call option with strike K and maturity T on an underlying asset with spot price S following a lognormal distribution. Show that, for any continuous function $g : \mathbb{R} \to \mathbb{R}$,

$$\lim_{x \searrow 0} \int_{-\infty}^{\infty} f(S)g(S)\, dS = g(K).$$

9. Consider the following first order ODE:

$$\begin{aligned} y'(x) &= y(x), \quad \forall\, x \in [0, 1]; \\ y(0) &= 1. \end{aligned}$$

(i) Discretize the interval $[0,1]$ using the nodes $x_i = ih$, $i = 0 : n$, where $h = \frac{1}{n}$. Use forward finite differences to obtain the following finite difference discretization of the ODE:

$$y_{i+1} = (1+h)y_i, \quad \forall\, i = 0 : (n-1),$$

with $y_0 = 1$. Show that

$$y_i = (1+h)^i, \quad \forall\, i = 0 : n.$$

(ii) Note that $y(x) = e^x$ it the exact solution of the ODE. Let

$$e_i = y_i - y(x_i) = (1+h)^i - e^{ih}$$

be the approximation error of the finite difference solution at the node x_i, $i = 0 : n$. Show that this finite difference discretization is convergent, i.e., that

$$\lim_{n\to\infty}\left(\max_{i=0:n}|e_i|\right) = 0.$$

Hint: It is easy to see that e_i can also be written as

$$e_i = e^{i\ln(1+h)} - e^{ih} = e^{ih}\left(e^{i(\ln(1+h)-h)} - 1\right).$$

Note that $ih \le 1$, for all $i = 0 : n$, since $h = \frac{1}{n}$.

Use the Taylor approximations $\ln(1+x) = x - \frac{x^2}{2} + O(x^3)$ and $e^x = 1 + x + O(x^2)$ to obtain that

$$i\ (\ln(1+h) - h) = -i\frac{h^2}{2} + O(h^3);$$
$$e^{i\ (\ln(1+h)-h)} - 1 = -i\frac{h^2}{2} + O(h^3).$$

Conclude that

$$\max_{i=0:n}|e_i| \le \frac{e}{2}\frac{1}{n} + O\left(\frac{1}{n^2}\right),$$

and therefore that

$$\lim_{n\to\infty}\left(\max_{i=0:n}|e_i|\right) = 0.$$

10. Consider the following second order ODE:

$$3x^2y''(x) - xy'(x) + y(x) = 0, \quad \forall\, x \in [0,1];$$
$$y(0) = 1; \qquad y(1) = \frac{1}{2}.$$

(i) Partition the interval $[0, 1]$ into n equal intervals, corresponding to nodes $x_i = ih$, $i = 0 : n$, where $h = \frac{1}{n}$. Write the finite difference discretization of the ODE at each node x_i, $i = 1 : (n-1)$, using central finite difference approximations for both $y'(x)$ and $y''(x)$.

(ii) If $n = 6$, we find, from the boundary conditions, that $y_0 = 1$ and $y_6 = \frac{1}{2}$. The finite difference discretization scheme presented above will have five equations can be written as a 5×5 linear system $AY = b$. Find A and b.

11. Show that the ODE

$$y''(x) - 2y'(x) + x^2 y(x) = 0$$

can be written as

$$Y'(x) = f(x, Y(x)),$$

where

$$Y(x) = \begin{pmatrix} y(x) \\ y'(x) \end{pmatrix} \quad \text{and} \quad f(x, Y(x)) = \begin{pmatrix} 0 & 1 \\ -x^2 & 2 \end{pmatrix} Y(x).$$

12. (i) Show that the approximate formula

$$1 + \frac{\sigma^2 S^2}{2} \cdot \frac{\Gamma}{\Theta} \approx 0$$

connecting the Gamma and the Theta of plain vanilla European options is exact if the underlying asset pays no dividends and if the risk–free interest rates are zero. In other words, for, e.g., call options, show that, if $r = q = 0$, then

$$1 + \frac{\sigma^2 S^2}{2} \cdot \frac{\Gamma(C)}{\Theta(C)} = 0.$$

(ii) If $q = 0$ but $r \neq 0$, show that

$$1 + \frac{\sigma^2 S^2}{2} \cdot \frac{\Gamma(C)}{\Theta(C)} = \frac{1}{1 + \frac{\sigma}{2r\sqrt{T-t}} \frac{N'(d_2)}{N(d_2)}}.$$

13. Show that the value of a plain vanilla European call option satisfies the Black–Scholes PDE. In other words, show that

$$\frac{\partial C}{\partial t} + \frac{1}{2}\sigma^2 S^2 \frac{\partial^2 C}{\partial S^2} + (r-q)S\frac{\partial C}{\partial S} - rC = 0,$$

where $C = C(S,t)$ is given by the Black–Scholes formula.

Hint: Although direct computation can be used to show this result, one could also use the version of the Black–Scholes PDE involving the Greeks, i.e.,

$$\Theta + \frac{1}{2}\sigma^2 S^2 \Gamma + (r-q)S\Delta - rV = 0,$$

and substitute the formulas for the Greeks.

14. The value at time t of a forward contract struck at K and maturing at time T, on an underlying asset with spot price S paying dividends continuously at the rate q, is

$$f(S,t) = Se^{-q(T-t)} - Ke^{-r(T-t)}.$$

Show that $f(S,t)$ satisfies the Black–Scholes PDE, i.e., show that

$$\frac{\partial f}{\partial t} + \frac{1}{2}\sigma^2 S^2 \frac{\partial^2 f}{\partial S^2} + (r-q)S\frac{\partial f}{\partial S} - rf = 0.$$

7.2 Solutions to Chapter 7 Exercises

Problem 1: Let $f(x) = x^3e^x - 6e^x$. Show that the central finite difference approximation for $f'(0)$ is a fourth order approximation.

Solution: Recall that, in general, the central finite difference approximation of the first derivative is a second order approximation, i.e.,

$$f'(0) \;=\; \frac{f(h) - f(-h)}{2h} \;+\; O\left(h^2\right), \quad \text{as } h \to 0. \tag{7.1}$$

To see why the central difference approximation for $f'(x)$ around the point 0 is a fourth order approximation for $f(x) = x^3e^x - 6e^x$, we investigate how the approximation (7.1) is derived.

The fifth order Taylor approximation of $f(x)$ around the point 0 is

$$\begin{aligned} f(x) \;=\;& f(0) + xf'(0) + \frac{x^2}{2}f''(0) + \frac{x^3}{6}f^{(3)}(0) + \frac{x^4}{24}f^{(4)}(0) + \frac{x^5}{120}f^{(5)}(0) \\ &+ O\left(x^6\right), \quad \text{as } x \to 0. \end{aligned} \tag{7.2}$$

We let $x = h$ and $x = -h$ in (7.2) and sum up the two resulting formulas. After solving for $f'(0)$ we obtain

$$f'(0) \;=\; \frac{f(h) - f(-h)}{2h} \;-\; \frac{h^2}{6}f^{(3)}(0) \;-\; \frac{h^4}{120}f^{(5)}(0) \;+\; O\left(h^5\right), \tag{7.3}$$

as $h \to 0$.

For $f(x) = x^3e^x - 6e^x$, we find that $f^{(3)}(x) = (x^3 + 9x^2 + 18x)e^x$, and thus that $f^{(3)}(0) = 0$. Also, $f^{(5)}(0) = 54 \neq 0$ and (7.3) becomes

$$f'(0) \;=\; \frac{f(h) - f(-h)}{2h} \;-\; \frac{9h^4}{20} \;+\; O\left(h^5\right) \;=\; \frac{f(h) - f(-h)}{2h} \;+\; O\left(h^4\right),$$

as $h \to 0$. In other words, the central difference approximation for $f'(x)$ around the point 0 is a fourth order approximation. □

Problem 2: Let $N(x)$ be the cumulative density of the standard normal variable. Show that the forward finite difference approximation of $N'(0)$ is a second order approximation.

Solution: The second order Taylor approximation of the function $N(x)$ around 0 can be written as

$$N(x) \;=\; N(0) \;+\; xN'(0) \;+\; \frac{x^2}{2}N''(0) \;+\; \frac{x^3}{6}N^{(3)}(0) \;+\; O(x^4), \tag{7.4}$$

as $x \to 0$. Recall that

$$N(x) = \frac{1}{\sqrt{2\pi}} \int_{-\infty}^{x} e^{-\frac{y^2}{2}}\, dy.$$

By differentiating the improper integral above with respect to x, we obtain that

$$N'(x) = \frac{1}{\sqrt{2\pi}} e^{-\frac{x^2}{2}},$$

and therefore that

$$N''(x) = -\frac{1}{\sqrt{2\pi}} x e^{-\frac{x^2}{2}};$$

$$N^{(3)}(x) = -\frac{1}{\sqrt{2\pi}} (1 - x^2) e^{-\frac{x^2}{2}}.$$

Thus,

$$N(0) = \frac{1}{2}; \quad N'(0) = \frac{1}{\sqrt{2\pi}}; \quad N''(0) = 0; \quad N^{(3)}(0) = -\frac{1}{\sqrt{2\pi}},$$

and the Taylor expansion (7.4) of $N(x)$ around the point 0 becomes[1]

$$N(x) = \frac{1}{2} + \frac{x}{\sqrt{2\pi}} - \frac{x^3}{6\sqrt{2\pi}} + O(x^4), \tag{7.5}$$

as $x \to 0$.

Then,

$$\frac{N(h) - N(0)}{h} = \frac{\left(\frac{1}{2} + \frac{h}{\sqrt{2\pi}} - \frac{h^3}{6\sqrt{2\pi}} + O(h^4)\right) - \frac{1}{2}}{h}$$

$$= \frac{1}{\sqrt{2\pi}} - \frac{h^2}{6\sqrt{2\pi}} + O(h^3)$$

$$= \frac{1}{\sqrt{2\pi}} + O(h^2)$$

$$= N'(0) + O(h^2),$$

as $h \to 0$.

[1]We note that the Taylor expansion of $N(x)$ can be more accurately written as

$$N(x) = \frac{1}{2} + \frac{x}{\sqrt{2\pi}} - \frac{x^3}{6\sqrt{2\pi}} + O(x^5),$$

as $x \to 0$, since $N^{(4)}(0) = 0$ and $N^{(5)}(0) = \frac{3}{\sqrt{2\pi}} \neq 0$. This level of precision is not needed here.

Note that, by definition, $-O(h^2)$ is the same as $O(h^2)$. We conclude that

$$N'(0) = \frac{N(h) - N(0)}{h} + O(h^2),$$

as $h \to 0$. In other words, the forward finite difference approximation of $N'(0)$ is a second order approximation, which is what we wanted to show. □

Problem 3: Find a second order finite difference approximation for $f'(a)$ using $f(a)$, $f(a+h)$, and $f(a+2h)$.

Solution: To obtain a finite difference approximation for $f'(a)$ in terms of $f(a)$, $f(a+h)$, and $f(a+2h)$ we use the cubic Taylor approximation of $f(x)$ around the point $x = a$, i.e.,

$$f(x) = f(a) + (x-a)f'(a) + \frac{(x-a)^2}{2}f''(a) + \frac{(x-a)^3}{6}f^{(3)}(a) + O\left((x-a)^4\right), \quad (7.6)$$

as $x \to a$. By letting $x = a + h$ and $x = a + 2h$ in (7.6), we obtain that

$$f(a+h) = f(a) + hf'(a) + \frac{h^2}{2}f''(a) + \frac{h^3}{6}f^{(3)}(a) + O\left(h^4\right); \quad (7.7)$$

$$f(a+2h) = f(a) + 2hf'(a) + 2h^2 f''(a) + \frac{4h^3}{3}f^{(3)}(a) + O\left(h^4\right), \quad (7.8)$$

as $h \to 0$.

We multiply (7.7) by 4 and subtract the result from (7.8) to obtain

$$f(a+2h) - 4f(a+h) = -3f(a) - 2hf'(a) + \frac{2h^3}{3}f^{(3)}(a) + O\left(h^4\right), \quad (7.9)$$

as $h \to 0$. By solving (7.9) for $f'(a)$, we obtain the following second order finite difference approximation of $f'(a)$:

$$\begin{aligned} f'(a) &= \frac{-f(a+2h) + 4f(a+h) - 3f(a)}{2h} + \frac{h^2}{3}f^{(3)}(a) + O\left(h^3\right) \\ &= \frac{-f(a+2h) + 4f(a+h) - 3f(a)}{2h} + O\left(h^2\right), \end{aligned}$$

as $h \to 0$. □

Problem 4: Let $f(x)$ be an infinitely many times differentiable function. Find a central finite difference approximation for the fourth derivative of f

at a, i.e., for $f^{(4)}(a)$, using $f(a-2h)$, $f(a-h)$, $f(a)$, $f(a+h)$, and $f(a+2h)$. What is the order of this finite difference approximation?

Solution: We will use the following Taylor approximation of $f(x)$ around the point $x = a$:

$$\begin{aligned} f(x) &= f(a) + (x-a)f'(a) + \frac{(x-a)^2}{2}f''(a) + \frac{(x-a)^3}{6}f^{(3)}(a) \\ &\quad + \frac{(x-a)^4}{24}f^{(4)}(a) + \frac{(x-a)^5}{120}f^{(5)}(a) + \frac{(x-a)^6}{720}f^{(6)}(a) \\ &\quad + O\left((x-a)^7\right), \end{aligned} \tag{7.10}$$

as $x \to a$.

For symmetry reasons, and keeping in mind the form of the central difference approximation for $f''(a)$, we use (7.10) to compute

$$\begin{aligned} f(a+h) + f(a-h) &= 2f(a) + h^2 f''(a) + \frac{h^4}{12}f^{(4)}(a) + \frac{h^6}{360}f^{(6)}(a) \\ &\quad + O\left(h^7\right), \quad \text{as } h \to 0; \end{aligned} \tag{7.11}$$

$$\begin{aligned} f(a+2h) + f(a-2h) &= 2f(a) + 4h^2 f''(a) + \frac{4h^4}{3}f^{(4)}(a) + \frac{8h^6}{45}f^{(6)}(a) \\ &\quad + O\left(h^7\right), \end{aligned} \tag{7.12}$$

as $h \to 0$. We multiply (7.11) by 4 and subtract the result from (7.12). We solve for $f^{(4)}(a)$ and obtain the following second order finite difference approximation:

$$f^{(4)}(a) = \frac{f(a+2h) - 4f(a+h) + 6f(a) - 4f(a-h) + f(a-2h)}{h^4} + O\left(h^2\right),$$

as $h \to 0$. □

Problem 5: The goal of this exercise is to emphasize the importance of symmetry in finite difference approximations. Recall that the central difference approximations for the first and second order derivatives are

$$\begin{aligned} f'(a) &= \frac{f(a+h) - f(a-h)}{2h} + O\left(h^2\right); \\ f''(a) &= \frac{f(a+h) - 2f(a) + f(a-h)}{h^2} + O\left(h^2\right), \end{aligned}$$

as $h \to 0$. In other words, $f'(a)$ and $f''(a)$ are approximated to second order accuracy by using the value of f at the point a and at the points $a-h$ and $a+h$ that are symmetric with respect to a.

We investigate what happens if symmetry is not required.

(i) Find a second order finite difference approximation of $f'(a)$ using $f(a)$, $f(a-h)$ and $f(a+2h)$.

(ii) Find a first order finite difference approximation of $f''(a)$ using $f(a)$, $f(a-h)$ and $f(a+2h)$. Note that, in general, a second order finite difference approximation of $f''(a)$ using $f(a)$, $f(a-h)$ and $f(a+2h)$ does not exist.

Let $\beta < a < \gamma$ such that $a - \beta = C(\gamma - a)$, where C is a constant.

(iii) Find a finite difference approximation of $f'(a)$ using $f(a)$, $f(\beta)$, and $f(\gamma)$ which is second order in terms of $|\gamma - a|$, i.e., where the residual term is $O\left(|\gamma - a|^2\right)$.

(iv) Find a finite difference approximation of $f''(a)$ using $f(a)$, $f(\beta)$, and $f(\gamma)$ which is first order in terms of $|\gamma - a|$. Show that, in general, a second order finite difference approximation of $f''(a)$ using $f(a)$, $f(\beta)$ and $f(\gamma)$ is not possible, unless $a = \frac{\beta+\gamma}{2}$, i.e., unless β and γ are symmetric with respect to a.

Solution: (i) and (ii). We use the cubic Taylor approximation of $f(x)$ around the point $x = a$, i.e.,

$$\begin{aligned} f(x) &= f(a) + (x-a)f'(a) + \frac{(x-a)^2}{2}f''(a) + \frac{(x-a)^3}{6}f^{(3)}(a) \\ &\quad + O\left((x-a)^4\right), \quad \text{as } x \to a. \end{aligned} \tag{7.13}$$

By letting $x = a - h$ and $x = a + 2h$ in (7.13), we obtain that

$$f(a-h) = f(a) - hf'(a) + \frac{h^2}{2}f''(a) - \frac{h^3}{6}f^{(3)}(a) + O\left(h^4\right); \tag{7.14}$$

$$f(a+2h) = f(a) + 2hf'(a) + 2h^2 f''(a) + \frac{4h^3}{3}f^{(3)}(a) + O\left(h^4\right), \tag{7.15}$$

as $h \to 0$.

We eliminate the terms containing $f''(a)$ by multiplying (7.14) by 4 and subtracting the result from (7.15). By solving for $f'(a)$, we obtain the following second order finite difference approximation of $f'(a)$:

$$\begin{aligned} f'(a) &= \frac{f(a+2h) + 3f(a) - 4f(a-h)}{6h} - \frac{h^2}{3}f^{(3)}(a) + O\left(h^3\right) \\ &= \frac{f(a+2h) + 3f(a) - 4f(a-h)}{6h} + O\left(h^2\right), \end{aligned}$$

as $h \to 0$.

Similarly, we eliminate the terms containing $f'(a)$ by multiplying (7.14) by 2 and adding the result to (7.15). By solving for $f''(a)$, we obtain the following first order finite difference approximation of $f''(a)$:

$$\begin{aligned} f''(a) &= \frac{f(a+2h) - 3f(a) + 2f(a-h)}{3h^2} - \frac{h}{3} f^{(3)}(a) + O\left(h^2\right) \\ &= \frac{f(a+2h) - 3f(a) + 2f(a-h)}{3h^2} + O(h), \end{aligned}$$

as $h \to 0$.

(iii) and (iv). Denote $\gamma - a$ by h, i.e., let $h = \gamma - a$. Then, $a - \beta = Ch$.

We write the cubic Taylor approximation (7.13) of $f(x)$ around the point a for $x = \gamma = a + h$ and for $x = \beta = a - Ch$ and obtain

$$f(\gamma) = f(a) + hf'(a) + \frac{h^2}{2} f''(a) + \frac{h^3}{6} f^{(3)}(a) + O\left(h^4\right); \quad (7.16)$$

$$f(\beta) = f(a) - Chf'(a) + \frac{C^2h^2}{2} f''(a) - \frac{C^3h^3}{6} f^{(3)}(a) + O\left(h^4\right), \quad (7.17)$$

as $h \to 0$.

By eliminating from (7.16) and (7.17) the terms containing $f''(a)$ and solving for $f'(a)$ we obtain the following finite difference approximation:

$$f'(a) = \frac{C^2 f(\gamma) - (C^2 - 1) f(a) - f(\beta)}{C(C+1)h} + O\left(h^2\right). \quad (7.18)$$

Similarly, we eliminate from (7.16) and (7.17) the terms containing $f'(a)$ and solve for $f''(a)$ to obtain the following finite difference approximation:

$$f''(a) = 2\frac{Cf(\gamma) - (C+1)f(a) + f(\beta)}{C(C+1)h^2} + O(h). \quad (7.19)$$

Note that, in general, the finite difference approximation (7.18) of $f'(a)$ is second order, while the finite difference approximation (7.19) of $f''(a)$ is first order. The finite difference approximation (7.19) of $f''(a)$ would be second order, e.g., if $C = 1$ or if $f^{(3)}(a) = 0$.

Also, note that, for $C = 1$, i.e., if $\beta = a - h$ and $\gamma = a + h$ are symmetric with respect to the point a, then (7.18) becomes the central finite difference approximation of $f'(a)$, i.e.,

$$f'(a) = \frac{f(a+h) - f(a-h)}{2h} + O\left(h^2\right).$$

The same would not be true for (7.19), which becomes

$$f''(a) = \frac{f(a+h) - 2f(a) + f(a-h)}{h^2} + O(h), \quad (7.20)$$

instead of the central finite difference approximation

$$f''(a) = \frac{f(a+h) - 2f(a) + f(a-h)}{h^2} + O\left(h^2\right)$$

of $f''(a)$. This is due to the fact that, for $C = 1$, the coefficient of h from $O(h)$ from (7.20) cancels out and the next term of order $O\left(h^2\right)$ becomes relevant. □

Problem 6: A butterfly spread is made of a long position in a call option with strike $K - x$, a long position in a call option with strike $K + x$, and a short position in two calls with strike K. The options are on the same underlying asset and have the same maturities.

(i) Show that the value of the butterfly spread is

$$C(K+x) - 2C(K) + C(K-x),$$

where, e.g., $C(K+x)$ denotes the price of the call with strike $K + x$.

(ii) Show that, in the limiting case when x goes to 0, the **value** of a position in $\frac{1}{x^2}$ butterfly spreads as above converges to the second order partial derivative of the value of the option, C, with respect to strike K, i.e., show that

$$\lim_{x \searrow 0} \frac{C(K+x) - 2C(K) + C(K-x)}{x^2} = \frac{\partial^2 C}{\partial K^2}(K).$$

(iii) Show that, in the limiting case when $x \to 0$, the **payoff at maturity** of a position in $\frac{1}{x}$ butterfly spreads as above is going to approximate the payoff of a derivative security that pays 1 if the underlying asset expires at K, and 0 otherwise.

Solution: (i) The value of a butterfly spread, i.e., of a long position in a call option with strike $K - x$ and value $C(K-x)$, a long position in a call option with strike $K + x$ and value $C(K+x)$, and a short position in two calls with strike K and value $-2C(K)$ is

$$C(K+x) - 2C(K) + C(K-x).$$

(ii) The value of a position in $\frac{1}{x^2}$ butterfly spreads is

$$\frac{1}{x^2}\left(C(K+x) - 2C(K) + C(K-x)\right). \tag{7.21}$$

The value of a call option as a function of the strike of the option is infinitely many times differentiable (for any fixed point in time except at maturity).

The expression from (7.21) represents the central finite difference approximation of $\frac{\partial^2 C}{\partial K^2}(K)$. We know that

$$\frac{C(K+x) - 2C(K) + C(K-x)}{x^2} = \frac{\partial^2 C}{\partial K^2}(K) + O(x^2).$$

Then,

$$\lim_{x \searrow 0} \frac{C(K+x) - 2C(K) + C(K-x)}{x^2} = \frac{\partial^2 C}{\partial K^2}(K).$$

(iii) The payoff at maturity of the butterfly spread is

$$\max(S-(K-x),0) - 2\max(S-K,0) + \max(S-(K+x),0)$$
$$= \begin{cases} 0, & \text{if } S < K-x; \\ S-(K-x), & \text{if } K-x \le S \le K; \\ K+x-S, & \text{if } K \le S \le K+x; \\ 0, & \text{if } K+x < S. \end{cases}$$

Denote by $f_x(S)$ the payoff at maturity of a position in $\frac{1}{x}$ butterfly spreads. Then,

$$f_x(S) = \begin{cases} 0, & \text{if } S < K-x; \\ \frac{S-(K-x)}{x}, & \text{if } K-x \le S \le K; \\ \frac{K+x-S}{x}, & \text{if } K \le S \le K+x; \\ 0, & \text{if } K+x < S. \end{cases}$$

Note that $f_x(K) = 1$ for any $x \neq 0$, and therefore

$$\lim_{x \searrow 0} f_x(K) = 1. \tag{7.22}$$

Let $S \neq K$ be a fixed value of the spot price of the underlying asset. Then $f_x(S) = 0$ for any x such that $0 < x < |K-S|$, and therefore

$$\lim_{x \searrow 0} f_x(S) = 0, \ \forall\, S \neq K. \tag{7.23}$$

From (7.22) and (7.23), we conclude that, in the limiting case when $x \to 0$, the payoff at maturity of a position in $\frac{1}{x}$ butterfly spreads as above is 1 if the underlying asset expires at K, and 0 otherwise. □

Problem 7: A bull spread is made of a long position in a call option with strike K and a short position in a call option with strike $K+x$, both options being on the same underlying asset and having the same maturities. Let $C(K)$ and $C(K+x)$ be the values (at time t) of the call options with strikes K and $K+x$, respectively.

(i) The **value** of a position in $\frac{1}{x}$ bull spreads is $\frac{C(K)-C(K+x)}{x}$. In the limiting case when x goes to 0, show that

$$\lim_{x\searrow 0} \frac{C(K)-C(K+x)}{x} = -\frac{\partial C}{\partial K}(K).$$

(ii) Show that, in the limiting case when $x \to 0$, the **payoff at maturity** of a position in $\frac{1}{x}$ bull spreads as above is going to approximate the payoff of a derivative security that pays 1 if the price of the underlying asset at expiry is above K, and pays 0 if the price of the underlying asset at expiry is below K.

Solution: (i) Since the value $C(K)$ of a call option as a function of the strike K of the option is infinitely many times differentiable, the first order forward finite difference approximation of $\frac{\partial C}{\partial K}(K)$ is

$$\frac{\partial C}{\partial K}(K) = \frac{C(K+x)-C(K)}{x} + O(x),$$

as $x \to 0$. We conclude that

$$\lim_{x\searrow 0} \frac{C(K)-C(K+x)}{x} = -\frac{\partial C}{\partial K}(K).$$

(ii) The payoff at maturity of the bull spread is

$$\max(S-K,0) - \max(S-(K+x),0) = \begin{cases} 0, & \text{if } S \leq K; \\ S-K, & \text{if } K < S \leq K+x; \\ x, & \text{if } K+x < S. \end{cases}$$

If $g_x(S)$ denotes the payoff at maturity of a position in $\frac{1}{x}$ bull spreads, then

$$g_x(S) = \begin{cases} 0, & \text{if } S \leq K; \\ \frac{S-K}{x}, & \text{if } K < S \leq K+x; \\ 1, & \text{if } K+x < S. \end{cases}$$

If $S \leq K$, then $g_x(S) = 0$ for any $x > 0$ and therefore

$$\lim_{x\searrow 0} g_x(S) = 0, \quad \forall\, S \leq K. \tag{7.24}$$

If $S > K$, then $g_x(S) = 1$ for any x such that $0 < x < S-K$, and therefore

$$\lim_{x\searrow 0} g_x(S) = 1, \quad \forall\, S > K. \tag{7.25}$$

From (7.24) and (7.25), we conclude that, in the limiting case when $x \to 0$, the payoff at maturity of a position in $\frac{1}{x}$ bull spreads as above is

1 if the underlying asset expires above K, and 0 if the underlying asset expires below K. $\square$

Problem 8: Let

$$f(S) = \frac{C(K+x,T) - 2C(K,T) + C(K-x,T)}{x^2},$$

where, e.g., $C(K,T)$ denotes the value of a plain vanilla call option with strike K and maturity T on an underlying asset with spot price S following a lognormal distribution. Show that, for any continuous function $g : \mathbb{R} \to \mathbb{R}$,

$$\lim_{x \searrow 0} \int_{-\infty}^{\infty} f(S)g(S)\, dS = g(K). \tag{7.26}$$

Solution: From the definition of $f(S)$, it is easy to see that

$$\begin{aligned} f(S) &= \frac{\max(S-(K-x),0) - 2\max(S-K,0) + \max(S-(K+x),0)}{x^2} \\ &= \begin{cases} 0, & \text{if } 0 < S \le K-x; \\ \frac{S-(K-x)}{x^2}, & \text{if } K-x < S \le K; \\ \frac{K+x-S}{x^2}, & \text{if } K < S \le K+x; \\ 0, & \text{if } K+x < S. \end{cases} \end{aligned}$$

Then,

$$\begin{aligned} \int_{-\infty}^{\infty} f(S)g(S)dS &= \int_{K-x}^{K} \frac{S-(K-x)}{x^2} g(S)dS + \int_{K}^{K+x} \frac{K+x-S}{x^2} g(S)dS \\ &= \frac{1}{x^2}\int_0^x zg(K-x+z)dz + \frac{1}{x^2}\int_0^x wg(K+x-w)dw, \end{aligned}$$

where we used the substitutions $z = S-(K-x)$ and $w = K+x-S$ for the two integrals above, respectively.

Let $z = xy$ and $w = xt$. Then $dz = x\,dy$, $dw = x\,dt$, and we find that

$$\begin{aligned} \int_{-\infty}^{\infty} f(S)g(S)\, dS &= \frac{1}{x^2}\int_0^x zg(K-x+z)dz + \frac{1}{x^2}\int_0^x wg(K+x-w)dw \\ &= \int_0^1 yg(K-x+xy)dy + \int_0^1 tg(K+x-xt)dt. \end{aligned} \tag{7.27}$$

We let $x \searrow 0$ in (7.27). Since the function $g : \mathbb{R} \to \mathbb{R}$ is continuous, we obtain that

$$\begin{aligned} \lim_{x \searrow 0} \int_{-\infty}^{\infty} f(S)g(S)\, dS &= g(K)\int_0^1 y\, dy \ + \ g(K)\int_0^1 t\, dt \\ &= g(K)\cdot\frac{1}{2} \ + \ g(K)\cdot\frac{1}{2} \\ &= g(K), \end{aligned} \tag{7.28}$$

which is what we wanted to show; cf. (7.26).

For the sake of completeness, we provide rigorous proof of the fact that (7.27) becomes (7.28) when $x \searrow 0$. To do so, it is enough to show that

$$\lim_{x \searrow 0} \int_0^1 yg(K - x + xy)\, dy \;=\; g(K) \int_0^1 y\, dy.$$

Let $\epsilon > 0$ arbitrary. Since g is continuous, it follows that there exists $\delta > 0$ such that $|g(K) - g(\tau)| < \epsilon$ for all τ such that $|K - \tau| < \delta$. Let $x \in (0, \delta)$ and $y \in (0, 1)$. Then $|K - (K - x + xy)| = x(1 - y) < \delta$ and therefore

$$|g(K) - g(K - x + xy)| < \epsilon, \quad \forall\, 0 < x < \delta,\ 0 < y < 1.$$

Therefore, it is easy to see that, for any $0 < x < \delta$,

$$\begin{aligned}
&\left| \int_0^1 yg(K - x + xy)dy \;-\; g(K) \int_0^1 ydy \right| \\
=\;& \left| \int_0^1 y\left(g(K - x + xy) - g(K)\right) dy \right| \\
\leq\;& \int_0^1 y|g(K) - g(K - x + xy)|dy \\
<\;& \epsilon \int_0^1 y\, dy \\
=\;& \frac{\epsilon}{2}.
\end{aligned}$$

In other words, for any $\epsilon > 0$ there exists $\delta > 0$ such that

$$\left| \int_0^1 yg(K - x + xy)\, dy \;-\; g(K) \int_0^1 y\, dy \right| \;<\; \frac{\epsilon}{2}, \quad \forall\, 0 < x < \delta.$$

Then, by definition,

$$\lim_{x \searrow 0} \left| \int_0^1 yg(K - x + xy)\, dy \;-\; g(K) \int_0^1 y\, dy \right| \;=\; 0. \quad \square$$

Problem 9: Consider the following first order ODE:

$$\begin{aligned}
y'(x) &= y(x), \quad \forall\, x \in [0, 1]; \\
y(0) &= 1.
\end{aligned}$$

(i) Discretize the interval $[0,1]$ using the nodes $x_i = ih$, $i = 0 : n$, where $h = \frac{1}{n}$. Use forward finite differences to obtain the following finite difference discretization of the ODE:

$$y_{i+1} = (1+h)y_i, \quad \forall\, i = 0 : (n-1),$$

with $y_0 = 1$. Show that

$$y_i = (1+h)^i, \quad \forall\, i = 0 : n.$$

(ii) Note that $y(x) = e^x$ is the exact solution of the ODE. Let

$$e_i = y_i - y(x_i) = (1+h)^i - e^{ih}$$

be the approximation error of the finite difference solution at the node x_i, $i = 0 : n$. Show that this finite difference discretization is convergent, i.e., that

$$\lim_{n\to\infty}\left(\max_{i=0:n}|e_i|\right) = 0.$$

Solution: (i) Recall that

$$y_0 = 1 \quad \text{and} \quad y_{i+1} = (1+h)y_i, \quad \forall\, i = 0 : (n-1).$$

We prove by induction that $y_i = (1+h)^i$, for all $i = 0 : n$, as follows:
Initial condition: for $i = 0$, we know that $y_0 = 1 = (1+h)^0$.
Induction step: assume that $y_i = (1+h)^i$. Then,

$$y_{i+1} = (1+h)y_i = (1+h)^{i+1},$$

which is what we wanted to show.
(ii) Let $y(x) = e^x$ be the exact solution of the ODE. It is easy to see that the approximation error e_i can also be written as

$$\begin{aligned} e_i &= y_i - y(x_i) = (1+h)^i - e^{ih} = e^{i\ln(1+h)} - e^{ih} \\ &= e^{ih}\left(e^{i(\ln(1+h)-h)} - 1\right). \end{aligned}$$

Using the Taylor approximation $\ln(1+x) = x - \frac{x^2}{2} + O(x^3)$ we find that

$$\begin{aligned} i\,(\ln(1+h) - h) &= i\left(\left(h - \frac{h^2}{2} + O(h^3)\right) - h\right) \\ &= -i\frac{h^2}{2} + iO(h^3) = -i\frac{h^2}{2} + O(h^2), \end{aligned}$$

since $ih \le 1$, for all $i = 0 : n$. Note that the estimate $iO(h^3) = O(h^2)$ is sharp, since, for $i = n$, the product ih is equal to $ih = nh = 1$.

From the Taylor approximation $e^x = 1 + x + O(x^2)$, it follows that

$$\begin{aligned} e^{i\,[\ln(1+h)-h]} - 1 &= \exp\left(-i\frac{h^2}{2} + O(h^2)\right) - 1 \\ &= 1 + \left(-i\frac{h^2}{2} + O(h^2)\right) + O\left(\left(-i\frac{h^2}{2} + O(h^2)\right)^2\right) - 1 \\ &= -i\frac{h^2}{2} + O(h^2), \end{aligned}$$

since $ih \leq 1$ for all $i = 0 : n$, and therefore

$$O\left(\left(-i\frac{h^2}{2} + O(h^2)\right)^2\right) = O\left(\left(-\frac{h}{2} + O(h^2)\right)^2\right) = O(h^2).$$

Since $e^{ih} \leq e$ for all $i = 0 : n$, we obtain that

$$\begin{aligned} \max_{i=0:n} |e_i| &= \max_{i=0:n} \left| e^{ih} \left(e^{i(\ln(1+h)-h)} - 1 \right) \right| \\ &\leq e \max_{i=0:n} \left| e^{i\,(\ln(1+h)-h)} - 1 \right| \\ &\leq e \max_{i=0:n} \left| -i\frac{h^2}{2} + O(h^2) \right| \\ &\leq e\,\frac{h}{2} + O(h^2) \\ &= O(h) = O\left(\frac{1}{n}\right). \end{aligned}$$

We conclude that

$$\lim_{n\to\infty} \left(\max_{i=0:n} |e_i| \right) = 0,$$

and therefore that the finite difference discretization scheme of the ODE is convergent.

Note that we actually showed that the finite difference discretization is first order convergent, i.e.,

$$\max_{i=0:n} |e_i| = O\left(\frac{1}{n}\right). \quad \square$$

Problem 10: Consider the following second order ODE:

$$3x^2 y''(x) - xy'(x) + y(x) = 0, \quad \forall\, x \in [0, 1]; \tag{7.29}$$

$$y(0) = 1; \qquad y(1) = \frac{1}{2}. \tag{7.30}$$

(i) Partition the interval $[0, 1]$ into n equal intervals, corresponding to nodes $x_i = ih$, $i = 0 : n$, where $h = \frac{1}{n}$. Write the finite difference discretization of the ODE at each node x_i, $i = 1 : (n-1)$, using central finite difference approximations for both $y'(x)$ and $y''(x)$.

(ii) If $n = 6$, we find, from the boundary conditions, that $y_0 = 1$ and $y_6 = \frac{1}{2}$. The finite difference discretization scheme presented above will have five equations can be written as a 5×5 linear system $AY = b$. Find A and b.

Solution: (i) Let $x_i = ih$, $i = 0 : n$, where $h = \frac{1}{n}$. We look for $y_0, y_1, \ldots, y_n$ such that y_i is an approximate value of $y(x_i)$, for all $i = 0 : n$.

By writing the ODE (7.29) at each interior node $x_i = ih$, $i = 1 : (n-1)$, we obtain that

$$3x_i^2 y''(x_i) - x_i y'(x_i) + y(x_i) \;=\; 0, \quad \forall\, i = 1 : (n-1). \tag{7.31}$$

We substitute the second order central difference approximations for $y''(x_i)$ and $y'(x_i)$, respectively, i.e.,

$$\begin{aligned} y''(x_i) &= \frac{y(x_{i+1}) \;-\; 2y(x_i) \;+\; y(x_{i-1})}{h^2} \;+\; O(h^2); \\ y'(x_i) &= \frac{y(x_{i+1}) \;-\; y(x_{i-1})}{2h} \;+\; O(h^2), \end{aligned}$$

into (7.31), use the approximate values y_i for the exact values $y(x_i)$, for $i = 0 : n$, and ignore the $O(h^2)$ term. The following second order finite difference discretization of (7.29) is obtained:

$$3i^2h^2\, \frac{y_{i+1} - 2y_i + y_{i-1}}{h^2} \;-\; ih\, \frac{y_{i+1} - y_{i-1}}{2h} \;+\; y_i \;=\; 0, \quad \forall\, i = 1 : (n-1),$$

since $x_i = ih$, which can be written as

$$\left(3i^2 + \frac{i}{2}\right) y_{i-1} \;-\; (6i^2 - 1)y_i \;+\; \left(3i^2 - \frac{i}{2}\right) y_{i+1} \;=\; 0, \quad \forall\, i = 1 : (n-1). \tag{7.32}$$

From the boundary conditions (7.30), we find that $y_0 = 1$ and $y_n = \frac{1}{2}$.

(ii) For $n = 6$, the finite difference discretization (7.32) of the ODE (7.29) can be written in matrix form as

$$A\,Y \;=\; b,$$

where A is a tridiagonal 5×5 matrix given by

$$\begin{aligned} A(i,i) &= -(6i^2 - 1), && \forall\; i = 1 : 5; \\ A(i,i-1) &= 3i^2 + \tfrac{i}{2}, && \forall\; i = 2 : 5; \\ A(i,i+1) &= 3i^2 - \tfrac{i}{2}, && \forall\; i = 1 : 4, \end{aligned}$$

i.e.,

$$A = \begin{pmatrix} -5 & 2.5 & 0 & 0 & 0 \\ 13 & -23 & 11 & 0 & 0 \\ 0 & 28.5 & -53 & 25.5 & 0 \\ 0 & 0 & 50 & -95 & 46 \\ 0 & 0 & 0 & 77.5 & -149 \end{pmatrix},$$

and Y and b are the following column vectors:

$$Y = \begin{pmatrix} y_1 \\ y_2 \\ y_3 \\ y_4 \\ y_5 \end{pmatrix}; \quad b = \begin{pmatrix} -3.5 \\ 0 \\ 0 \\ 0 \\ -36.25 \end{pmatrix}. \quad \square$$

Problem 11: Show that the ODE

$$y''(x) - 2y'(x) + x^2 y(x) = 0$$

can be written as

$$Y'(x) = f(x, Y(x)),$$

where

$$Y(x) = \begin{pmatrix} y(x) \\ y'(x) \end{pmatrix} \quad \text{and} \quad f(x, Y(x)) = \begin{pmatrix} 0 & 1 \\ -x^2 & 2 \end{pmatrix} Y(x).$$

Solution: Note that $y''(x) = 2y'(x) - x^2 y(x)$. Then,

$$\begin{aligned} Y'(x) &= \begin{pmatrix} y'(x) \\ y''(x) \end{pmatrix} = \begin{pmatrix} y'(x) \\ 2y'(x) - x^2 y(x) \end{pmatrix} \\ &= \begin{pmatrix} 0 & 1 \\ -x^2 & 2 \end{pmatrix} \begin{pmatrix} y(x) \\ y'(x) \end{pmatrix} = \begin{pmatrix} 0 & 1 \\ -x^2 & 2 \end{pmatrix} Y(x). \quad \square \end{aligned}$$

Problem 12: (i) Show that the approximate formula

$$1 + \frac{\sigma^2 S^2}{2} \cdot \frac{\Gamma}{\Theta} \approx 0$$

connecting the Gamma and the Theta of plain vanilla European options is exact if the underlying asset pays no dividends and if the risk–free interest rates are zero. In other words, for, e.g., call options[2], show that

$$1 + \frac{\sigma^2 S^2}{2} \cdot \frac{\Gamma(C)}{\Theta(C)} = 0,$$

[2]Note that $\Gamma(P) = \Gamma(C)$, and, if $r = q = 0$, then $\Theta(P) = \Theta(C)$.

if $q = 0$ and $r = 0$.

(ii) If $q = 0$ and $r \neq 0$, show that

$$1 + \frac{\sigma^2 S^2}{2} \cdot \frac{\Gamma(C)}{\Theta(C)} = \frac{1}{1 + \frac{\sigma}{2r\sqrt{T}} \frac{N'(d_2)}{N(d_2)}}.$$

Solution: Recall that the Gamma and the Theta of plain vanilla European options are

$$\Gamma(C) = \frac{e^{-qT}}{\sigma S\sqrt{2\pi T}} e^{-\frac{d_1^2}{2}}; \tag{7.33}$$

$$\Theta(C) = -\frac{\sigma S e^{-qT}}{2\sqrt{2\pi T}} e^{-\frac{d_1^2}{2}} + qSe^{-qT}N(d_1) - rKe^{-rT}N(d_2); \tag{7.34}$$

$$\Gamma(P) = \frac{e^{-qT}}{\sigma S\sqrt{2\pi T}} e^{-\frac{d_1^2}{2}}; \tag{7.35}$$

$$\Theta(P) = -\frac{S\sigma e^{-qT}}{2\sqrt{2\pi T}} e^{-\frac{d_1^2}{2}} - qSe^{-qT}N(-d_1) + rKe^{-rT}N(-d_2), \tag{7.36}$$

where $d_1 = \left(\ln\left(\frac{S}{K}\right) + \left(r + \frac{\sigma^2}{2}\right)T\right) / \left(\sigma\sqrt{T}\right)$ and $d_2 = d_1 - \sigma\sqrt{T}$.

(i) For $r = q = 0$, we obtain from (7.33) and (7.34) that

$$\Gamma(C) = \frac{1}{\sigma S\sqrt{2\pi T}} e^{-\frac{d_1^2}{2}}; \quad \Theta(C) = -\frac{\sigma S}{2\sqrt{2\pi T}} e^{-\frac{d_1^2}{2}}.$$

Then,

$$1 + \frac{\sigma^2 S^2}{2} \cdot \frac{\Gamma(C)}{\Theta(C)} = 1 + \frac{\sigma^2 S^2}{2}\left(-\frac{2}{\sigma^2 S^2}\right) = 0. \tag{7.37}$$

Recall that $\Gamma(P) = \Gamma(C)$; cf. (7.33) and (7.35). From (7.34) and (7.36), it is easy to see that, if $r = q = 0$, then $\Theta(P) = \Theta(C)$. Thus, from (7.37), it follows that

$$1 + \frac{\sigma^2 S^2}{2} \cdot \frac{\Gamma(P)}{\Theta(P)} = 0.$$

(ii) For $q = 0$, we obtain from (7.33) and (7.34) that

$$\Gamma(C) = \frac{1}{\sigma S\sqrt{2\pi T}} e^{-\frac{d_1^2}{2}};$$

$$\Theta(C) = -\frac{\sigma S}{2\sqrt{2\pi T}} e^{-\frac{d_1^2}{2}} - rKe^{-rT}N(d_2).$$

Then,

$$
\begin{aligned}
1+\frac{\sigma^2 S^2}{2}\cdot\frac{\Gamma(C)}{\Theta(C)} &= 1-\frac{\frac{\sigma S}{2\sqrt{2\pi T}}e^{-\frac{d_1^2}{2}}}{\frac{\sigma S}{2\sqrt{2\pi T}}e^{-\frac{d_1^2}{2}} + rKe^{-rT}N(d_2)}\\
&= \frac{rKe^{-rT}N(d_2)}{\frac{\sigma S}{2\sqrt{2\pi T}}e^{-\frac{d_1^2}{2}} + rKe^{-rT}N(d_2)}\\
&= \frac{1}{1+\frac{\sigma}{2r\sqrt{T}}\cdot\frac{S}{Ke^{-rT}N(d_2)}\cdot\frac{1}{\sqrt{2\pi}}e^{-\frac{d_1^2}{2}}}\\
&= \frac{1}{1+\frac{\sigma}{2r\sqrt{T}}\cdot\frac{SN'(d_1)}{Ke^{-rT}N(d_2)}}. \qquad (7.38)
\end{aligned}
$$

since $N'(t) = \frac{1}{\sqrt{2\pi}}e^{-\frac{t^2}{2}}$ for all $t \in \mathbb{R}$.

Recall that the "magic" of Greek computations is due to the following result:

$$S\ N'(d_1) = Ke^{-rT}\ N'(d_2);$$

cf. Lemma 3.15 of [1] for $q = 0$. Then, (7.38) becomes

$$1+\frac{\sigma^2 S^2}{2}\cdot\frac{\Gamma(C)}{\Theta(C)} = \frac{1}{1+\frac{\sigma}{2r\sqrt{T}}\frac{N'(d_2)}{N(d_2)}}. \quad \square$$

Problem 13: Show that the value of a plain vanilla European call option satisfies the Black–Scholes PDE. In other words, show that

$$\frac{\partial C}{\partial t} + \frac{1}{2}\sigma^2 S^2\frac{\partial^2 C}{\partial S^2} + (r-q)S\frac{\partial C}{\partial S} - rC = 0,$$

where $C = C(S,t)$ is given by the Black–Scholes formula.

Solution: Although direct computation can be used to show this result, we will use the version of the Black–Scholes PDE involving the Greeks, i.e.,

$$\Theta + \frac{1}{2}\sigma^2 S^2\Gamma + (r-q)S\Delta - rC = 0,$$

substitute for Δ, Γ and Θ the values

$$
\begin{aligned}
\Delta &= e^{-q(T-t)}N(d_1);\\
\Gamma &= \frac{e^{-q(T-t)}}{S\sigma\sqrt{2\pi(T-t)}}e^{-\frac{d_1^2}{2}};\\
\Theta &= qSe^{-q(T-t)}N(d_1) - rKe^{-r(T-t)}N(d_2) - \frac{\sigma Se^{-q(T-t)}}{2\sqrt{2\pi(T-t)}}e^{-\frac{d_1^2}{2}},
\end{aligned}
$$

and substitute for C the value given by the Black–Scholes formula, i.e.,

$$C = Se^{-q(T-t)}N(d_1) - Ke^{-r(T-t)}N(d_2).$$

Then,

$$\begin{aligned}
& \Theta + \frac{1}{2}\sigma^2 S^2\Gamma + (r-q)S\Delta - rC \\
= \ & qSe^{-q(T-t)}N(d_1) - rKe^{-r(T-t)}N(d_2) - \frac{\sigma Se^{-q(T-t)}}{2\sqrt{2\pi(T-t)}}e^{-\frac{d_1^2}{2}} \\
& + \frac{\sigma^2 S^2}{2}\frac{e^{-q(T-t)}}{S\sigma\sqrt{2\pi(T-t)}}e^{-\frac{d_1^2}{2}} \\
& + (r-q)Se^{-q(T-t)}N(d_1) - r\left(Se^{-q(T-t)}N(d_1) - Ke^{-r(T-t)}N(d_2)\right) \\
= \ & \left(qSe^{-q(T-t)} + (r-q)Se^{-q(T-t)} - rSe^{-q(T-t)}\right)N(d_1) \\
& + \left(-rKe^{-r(T-t)} + rKe^{-r(T-t)}\right)N(d_2) \\
& - \frac{\sigma Se^{-q(T-t)}}{2\sqrt{2\pi(T-t)}}e^{-\frac{d_1^2}{2}} + \frac{\sigma Se^{-q(T-t)}}{2\sqrt{2\pi(T-t)}}e^{-\frac{d_1^2}{2}} \\
= \ & 0. \quad \square
\end{aligned}$$

Problem 14: The value at time t of a forward contract struck at K and maturing at time T, on an underlying asset with spot price S paying dividends continuously at the rate q, is

$$f(S,t) = Se^{-q(T-t)} - Ke^{-r(T-t)}.$$

Show that $f(S,t)$ satisfies the Black–Scholes PDE, i.e., show that

$$\frac{\partial f}{\partial t} + \frac{1}{2}\sigma^2 S^2\frac{\partial^2 f}{\partial S^2} + (r-q)S\frac{\partial f}{\partial S} - rf = 0.$$

Solution: It is easy to see that

$$\begin{aligned}
\frac{\partial f}{\partial t} &= qSe^{-q(T-t)} - rKe^{-r(T-t)}; \\
\frac{\partial f}{\partial S} &= e^{-q(T-t)}; \\
\frac{\partial^2 f}{\partial S^2} &= 0.
\end{aligned}$$

Then,

$$\begin{aligned}
& \frac{\partial f}{\partial t} + \frac{1}{2}\sigma^2 S^2 \frac{\partial^2 f}{\partial S^2} + (r-q)S\frac{\partial f}{\partial S} - rf \\
= \; & qSe^{-q(T-t)} - rKe^{-r(T-t)} + 0 + (r-q)Se^{-q(T-t)} \\
& - r\left(Se^{-q(T-t)} - Ke^{-r(T-t)}\right) \\
= \; & 0. \quad \square
\end{aligned}$$

Chapter 8

Multivariable calculus. Extremum points. Barrier options. Optimality of early exercise.

8.1 Exercises

1. For $q = 0$, the formula for the Gamma of a plain vanilla European call option reduces to

$$\Gamma = \frac{1}{S\sigma\sqrt{2\pi T}} \exp\left(-\frac{d_1^2}{2}\right),$$

where

$$d_1 = \frac{\ln\left(\frac{S}{K}\right) + \left(r + \frac{\sigma^2}{2}\right)T}{\sigma\sqrt{T}}.$$

Show that, as a function of $S > 0$, the Gamma of the call option increases until it reaches a maximum point and then decreases. Also, show that

$$\lim_{S \searrow 0} \Gamma(S) = 0 \quad \text{and} \quad \lim_{S \to \infty} \Gamma(S) = 0.$$

2. Let D be the domain bounded by the x–axis, the y–axis, and the line $x + y = 1$. Compute

$$\int\int_D \frac{x-y}{x+y}\, dxdy.$$

Hint: Use the change of variables $s = x + y$ and $t = x - y$, which is equivalent to

$$x = \frac{s+t}{2} \quad \text{and} \quad y = \frac{s-t}{2}.$$

Note that $(x, y) \in D$ if and only if $0 \leq s \leq 1$ and $-s \leq t \leq s$.

3. Compute the integral of the function $f(x, y) = x^2 - 2y$ on the region bounded by the parabola $y = (x + 1)^2$ and the line $y = 5x - 1$.

4. Use the change of variables to polar coordinates to show that the area of a circle of radius R is πR^2, i.e., prove that

$$\int\int_{D(0,R)} 1 \; dxdy \;=\; \pi R^2.$$

5. Compute the area of the ellipse with semi axes a and b, i.e., compute

$$\int\int_D 1 \; dx \; dy,$$

where

$$D \;=\; \left\{ (x, y) \mid \frac{x^2}{a^2} + \frac{y^2}{b^2} \leq 1 \right\},$$

with $a, b > 0$, using the change of variables $x = ar\cos\theta$ and $y = br\sin\theta$.

6. Which number is larger, e^π or π^e?

7. Show that

$$\frac{e^x + e^y}{2} \;\geq\; e^{\frac{x+y}{2}}, \quad \forall\, x, y \in \mathbb{R}.$$

8. Let $u, v : [0, \infty) \to [0, \infty)$ be two continuous functions with positive values. Assume that there exists a constant $M > 0$ such that

$$u(x) \;\leq\; M + \int_0^x u(t)v(t)\; dt, \quad \forall\, x \geq 0.$$

Show that

$$u(x) \;\leq\; M \exp\left(\int_0^x v(t)\; dt \right), \quad \forall\, x \geq 0.$$

Hint: Investigate the monotonicity of the function

$$\left(M + \int_0^x u(t)v(t)\; dt \right) \exp\left(-\int_0^x v(t)\; dt \right).$$

Note: This is a version of Gronwall's inequality, and it is needed, e.g., to prove the uniqueness of the solution of an initial value problem for ordinary differential equations.

9. Let $V(S,t) = \exp(-ax - b\tau)u(x,\tau)$, where

$$x = \ln\left(\frac{S}{K}\right); \quad \tau = \frac{(T-t)\sigma^2}{2};$$

$$a = \frac{r-q}{\sigma^2} - \frac{1}{2}; \quad b = \left(\frac{r-q}{\sigma^2} + \frac{1}{2}\right)^2 + \frac{2q}{\sigma^2}$$

This is the change of variables that reduces the Black–Scholes PDE for $V(S,t)$ to the heat equation for $u(x,\tau)$.

(i) Show that the boundary condition $V(S,T) = \max(S-K,0)$ for the European call option becomes the following boundary condition for $u(x,\tau)$ at time $\tau = 0$:

$$u(x,0) = K \exp(ax) \max(e^x - 1, 0).$$

(ii) Show that the boundary condition $V(S,T) = \max(K-S,0)$ for the European put option becomes

$$u(x,0) = K \exp(ax) \max(1 - e^x, 0).$$

10. Show that the solution of the system

$$\begin{cases} 2a + 1 - \frac{2(r-q)}{\sigma^2} = 0 \\ b + a^2 + a\left(1 - \frac{2(r-q)}{\sigma^2}\right) - \frac{2r}{\sigma^2} = 0 \end{cases}$$

is

$$a = \frac{r-q}{\sigma^2} - \frac{1}{2}; \quad b = \left(\frac{r-q}{\sigma^2} + \frac{1}{2}\right)^2.$$

11. What does the boundary condition $V(B,t) = R$ for a down–and–out call with barrier B and rebate $R > 0$ correspond to for the function $u(x,\tau)$ given by

$$V(S,t) = \exp(-ax - b\tau)u(x,\tau),$$

where

$$x = \ln\left(\frac{S}{K}\right); \quad \tau = \frac{(T-t)\sigma^2}{2};$$

$$a = \frac{r-q}{\sigma^2} - \frac{1}{2}, \quad b = \left(\frac{r-q}{\sigma^2} + \frac{1}{2}\right)^2 + \frac{2q}{\sigma^2}.$$

12. Assume that the function $V(S, I, t)$ satisfies the PDE

$$\frac{\partial V}{\partial t} + \ln S \frac{\partial V}{\partial I} + \frac{1}{2}\sigma^2 S^2 \frac{\partial^2 V}{\partial S^2} + rS \frac{\partial V}{\partial S} - rV = 0. \tag{8.1}$$

Consider the following change of variables:

$$V(S, I, t) = F(y, t),$$

where

$$y = \frac{I + (T - t)\ln S}{T}.$$

Show that $F(y, t)$ satisfies the following PDE:

$$\frac{\partial F}{\partial t} + \frac{\sigma^2 (T-t)^2}{2T^2}\frac{\partial^2 F}{\partial y^2} + \left(r - \frac{\sigma^2}{2}\right)\frac{T-t}{T}\frac{\partial F}{\partial y} - rF = 0.$$

Note: The values of Asian options with continuously sampled geometric average satisfy the PDE (8.1).

13. Assume that the function $V(S, I, t)$ satisfies the following PDE:

$$\frac{\partial V}{\partial t} + S \frac{\partial V}{\partial I} + \frac{1}{2}\sigma^2 S^2 \frac{\partial^2 V}{\partial S^2} + rS \frac{\partial V}{\partial S} - rV = 0. \tag{8.2}$$

Consider the following change of variables:

$$V(S, I, t) = S\ H(R, t), \quad \text{where} \quad R = \frac{I}{S}. \tag{8.3}$$

Show that $H(R, t)$ satisfies the following PDE:

$$\frac{\partial H}{\partial t} + \frac{1}{2}\sigma^2 R^2 \frac{\partial^2 H}{\partial R^2} + (1 - rR)\frac{\partial H}{\partial R} = 0. \tag{8.4}$$

Note: The values of average strike Asian call options with continuously compounded arithmetic average satisfy the PDE (8.2). Similarity solutions of the type (8.3) are good candidates for solving the PDE (8.2). The PDE (8.4) satisfied by $H(R, T)$ can be solved numerically, e.g., by using finite differences.

14. The price of a non-dividend-paying asset is lognormally distributed. Assume that the spot price is 40, the volatility is 30%, and the interest rates are constant at 5%. Fill in the Black–Scholes values of the ITM put options on the asset in the table below:

Option Type	Strike	Maturity	Value
Put	45	6 months	
Put	45	3 months	
Put	48	6 months	
Put	48	3 months	
Put	51	6 months	
Put	51	3 months	

For which of these options is the intrinsic value $\max(K - S, 0)$ larger than the Black–Scholes value of the option (in which case the corresponding American put is guaranteed to be worth more than the European put)?

15. Show that the premium of the Black–Scholes value of a European call option over its intrinsic value $\max(S - K, 0)$ is largest at the money. In other words, show that the maximum value of

$$C_{BS}(S) - \max(S - K, 0)$$

is obtained for $S = K$, where $C_{BS}(S)$ is the Black–Scholes value of the plain vanilla European call option with strike K and spot price S.

Note: The premium of the Black–Scholes value of a European call option over its intrinsic value is called the time value of the option.

16. Use the formula

$$V(S, K, t) \;=\; C(S, K, t) \;-\; \left(\frac{B}{S}\right)^{2a} C\left(\frac{B^2}{S}, K, t\right), \qquad (8.5)$$

where

$$a = \frac{r - q}{\sigma^2} - \frac{1}{2}$$

, to find the value of a six months down–and–out call on a non–dividend–paying asset with price following a lognormal distribution with 30% volatility and spot price 40. The barrier is $B = 35$ and the strike for the call is $K = 40$. The risk–free interest rate is constant at 5%.

17. Show that the value of a down–and–out call with barrier B less than the strike K of the call, i.e., $B < K$, converges to the value of a plain vanilla call with strike K when $B \searrow 0$. For simplicity, assume that the underlying asset does not pay dividends and that interest rates are zero.

Hint: Show that the value of the down–and–out call is

$$V(S) \ = \ C(S) \ - \ \left(BN(d_1) \ - \ \frac{SK}{B} e^{-rT} \ N(d_2) \right),$$

where $C(S)$ is the value at time 0 of the plain vanilla call with strike K and

$$d_1 \ = \ \frac{\ln\left(\frac{B^2}{SK}\right) + \frac{\sigma^2}{2}T}{\sigma\sqrt{T}} \quad \text{and} \quad d_2 \ = \ \frac{\ln\left(\frac{B^2}{SK}\right) - \frac{\sigma^2}{2}T}{\sigma\sqrt{T}}.$$

Use l'Hôpital's rule to show that

$$\lim_{B \searrow 0} \frac{1}{B} \ N(d_2) \ = \ 0.$$

18. Compute the Delta and Gamma of a down–and–out call with $B < K$.

19. (i) Show that the value of a plain vanilla European call option on a non–dividend–paying asset is always greater than the intrinsic premium $\max(S - K, 0)$.

    (ii) Conclude that plain vanilla American call options on non–dividend–paying assets are never optimal to exercise early.

20. Show that the Black–Scholes value of a European call option on an asset paying dividends continuously at the rate $q > 0$ is smaller than the intrinsic value $S - K$ is the spot price S is large enough, regardless of how small q is.

21. For the same maturity, options with different strikes trade simultaneously. The goal of this problem is to compute the rate of change of the implied volatility as a function of the strikes of the options.

    In other words, assume that S, T, q and r are given, and let $C(K)$ be the (known) value of a call option with maturity T and strike K. Assume that options with all strikes K exist. Define the implied volatility $\sigma_{imp}(K)$ as the unique solution to

    $$C(K) \ = \ C_{BS}(K, \sigma_{imp}(K)),$$

    where $C_{BS}(K, \sigma_{imp}(K)) \ = \ C_{BS}(S, K, T, \sigma_{imp}(K), r, q)$ represents the Black–Scholes value of a call option with strike K on an underlying asset following a lognormal model with volatility $\sigma_{imp}(K)$. Find an implicit differential equation satisfied by $\sigma_{imp}(K)$, i.e., find

$$\frac{\partial \sigma_{imp}(K)}{\partial K}$$

as a function of $\sigma_{imp}(K)$.

8.2 Solutions to Chapter 8 Exercises

Problem 1: For $q = 0$, the formula for the Gamma of a plain vanilla European call option reduces to

$$\Gamma = \frac{1}{S\sigma\sqrt{2\pi T}} \exp\left(-\frac{(d_1(S))^2}{2}\right), \tag{8.6}$$

where

$$d_1(S) = \frac{\ln\left(\frac{S}{K}\right) + \left(r + \frac{\sigma^2}{2}\right)T}{\sigma\sqrt{T}}. \tag{8.7}$$

Show that, as a function of $S > 0$, the Gamma of the call option increases until it reaches a maximum point and then decreases. Also, show that

$$\lim_{S \searrow 0} \Gamma(S) = 0 \quad \text{and} \quad \lim_{S \to \infty} \Gamma(S) = 0. \tag{8.8}$$

Solution: From (8.6) we find that Γ can be written as

$$\Gamma(S) = \frac{1}{\sigma\sqrt{2\pi T}} \exp\left(-\frac{(d_1(S))^2}{2} - \ln(S)\right), \tag{8.9}$$

where $d_1(S)$ is given by (8.7).

Since $\Gamma(S) > 0$, it follows that the functions $\Gamma(S)$ and $\ln(\Gamma(S))$ have the same monotonicity intervals. Let $f : (0, \infty) \to \mathbb{R}$ given by

$$f(S) = \ln(\Gamma(S)) = -\frac{(d_1(S))^2}{2} - \ln(S) - \ln(\sigma\sqrt{2\pi T}).$$

Then,

$$\begin{aligned} f'(S) &= -d_1(S)\frac{\partial(d_1(S))}{\partial S} - \frac{1}{S} = -\frac{d_1(S)}{S\sigma\sqrt{T}} - \frac{1}{S} \\ &= -\frac{1}{S}\left(1 + \frac{d_1(S)}{\sigma\sqrt{T}}\right). \end{aligned} \tag{8.10}$$

Recall from (8.7) that

$$d_1(S) = \frac{\ln\left(\frac{S}{K}\right) + \left(r + \frac{\sigma^2}{2}\right)T}{\sigma\sqrt{T}}.$$

It is easy to see that $d_1(S)$ is an increasing function of S and that

$$\lim_{S \searrow 0} d_1(S) = -\infty; \quad \lim_{S \to \infty} d_1(S) = \infty. \tag{8.11}$$

From (8.10) we find that $f(S)$ has one critical point, denoted by S^*, with $d_1(S^*) = -\sigma\sqrt{T}$. From (8.10) and (8.11) it follows that $f'(S) > 0$ if $0 < S < S^*$ and $f'(S) < 0$ if $S^* < S$.

In other words, the function $f(S) = \ln(\Gamma(S))$ is increasing when $0 < S < S^*$ and is decreasing when $S^* < S$. We conclude that $\Gamma(S)$ is also increasing when $0 < S < S^*$ and decreasing when $S^* < S$.

We now compute $\lim_{S\searrow 0}\Gamma(S)$ and $\lim_{S\to\infty}\Gamma(S)$.

Note that $\lim_{S\to\infty} d_1(S) = \infty$. Therefore,

$$\lim_{S\to\infty}\Gamma(S) = \lim_{S\to\infty}\frac{1}{\sigma\sqrt{2\pi T}}\exp\left(-\frac{(d_1(S))^2}{2} - \ln(S)\right) = 0.$$

From (8.7), and using the fact that $\lim_{S\searrow 0}\ln(S) = -\infty$, it follows that

$$\begin{aligned}\lim_{S\searrow 0}\frac{-\frac{(d_1(S))^2}{2} - \ln(S)}{-(\ln(S))^2} &= \lim_{S\searrow 0}\left(\frac{\left(\ln(S) - \ln(K) + \left(r + \frac{\sigma^2}{2}\right)T\right)^2}{2\sigma^2 T(\ln(S))^2} + \frac{1}{\ln(S)}\right)\\ &= \frac{1}{2\sigma^2 T}.\end{aligned}$$

Since $\lim_{S\searrow 0}\left(\exp\left(-(\ln(S))^2\right)\right) = 0$, we obtain that

$$\lim_{S\searrow 0}\exp\left(\frac{-(d_1(S))^2}{2} - \ln(S)\right) = 0.$$

from (8.9), we conclude that

$$\lim_{S\searrow 0}\Gamma(S) = \lim_{S\searrow 0}\frac{1}{\sigma\sqrt{2\pi T}}\exp\left(-\frac{(d_1(S))^2}{2} - \ln(S)\right) = 0. \quad \square$$

Problem 2: Let D be the domain bounded by the x–axis, the y–axis, and the line $x + y = 1$. Compute

$$\int\int_D \frac{x-y}{x+y}\,dxdy. \tag{8.12}$$

Solution: Note that

$$D = \{(x,y) \mid x \geq 0, y \geq 0, x + y \leq 1\}.$$

We use the change of variables $s = x + y$ and $t = x - y$, which is equivalent to

$$x = \frac{s+t}{2}; \quad y = \frac{s-t}{2}.$$

It is easy to see that $(x, y) \in D$ if and only if $(s, t) \in \Omega$, where

$$\Omega = \{(s, t) \mid 0 \le s \le 1, -s \le t \le s\}.$$

The Jacobian of the change of variable $(x, y) \in D \longrightarrow (s, t) \in \Omega$ is

$$dxdy = \left| \frac{\partial x}{\partial s}\frac{\partial y}{\partial t} - \frac{\partial x}{\partial t}\frac{\partial y}{\partial s} \right| dsdt = \frac{1}{2} dsdt,$$

and therefore

$$\begin{aligned}
\int\int_D \frac{x-y}{x+y}\, dxdy &= \int\int_\Omega \frac{t}{s}\frac{1}{2}\, dsdt = \int_0^1 \left(\int_{-s}^s \frac{t}{2s}\, dt \right) ds \\
&= \frac{1}{2}\int_0^1 \frac{1}{s} \left(\int_{-s}^s t\, dt \right) ds = 0.
\end{aligned}$$

The integral (8.12) can also be estimated directly as follows:

$$\begin{aligned}
\int\int_D \frac{x-y}{x+y}\, dxdy &= \int_0^1 \left(\int_0^{1-y} \frac{x-y}{x+y}\, dx \right) dy \\
&= \int_0^1 \left((x - 2y\ln(x+y))\Big|_{x=0}^{x=1-y}\, dy \right) \\
&= \int_0^1 1 - y + 2y\ln(y)\, dy \\
&= \left(y - y^2 + y^2\ln(y) \right)\Big|_{y=0}^{y=1} \\
&= -\left(\lim_{y \searrow 0} y^2 \ln(y) \right).
\end{aligned}$$

Note that

$$\lim_{y\searrow 0} y^2\ln(y) = \lim_{t\to\infty} \left(\frac{1}{t}\right)^2 \ln\left(\frac{1}{t}\right) = -\lim_{t\to\infty} \frac{\ln(t)}{t^2} = 0.$$

We conclude that

$$\int\int_D \frac{x-y}{x+y}\, dxdy = 0. \quad \square$$

Problem 3: Compute the integral of the function $f(x, y) = x^2 - 2y$ on the region bounded by the parabola $y = (x+1)^2$ and the line $y = 5x - 1$.

Solution: We first identify the integration domain D. Note that $(x+1)^2 = 5x - 1$ if and only if $x = 1$ and $x = 2$, and that $(x+1)^2 \le 5x - 1$ if $1 < x < 2$. Therefore,

$$D = \{(x, y) \mid 1 \le x \le 2 \text{ and } (x+1)^2 \le y \le 5x - 1\}.$$

Then,

$$\begin{aligned}
\int\int_D f(x,y)dxdy &= \int_1^2 \left(\int_{(x+1)^2}^{5x-1} (x^2-2y)dy\right) dx \\
&= \int_1^2 \left((x^2y - y^2)\Big|_{y=(x+1)^2}^{y=5x-1} \right) dx \\
&= \int_1^2 x^2(5x-1-(x+1)^2) - ((5x-1)^2-(x+1)^4)dx \\
&= \int_1^2 (5x-1-(x+1)^2)(x^2-(5x-1+(x+1)^2))\ dx \\
&= \int_1^2 (-x^2+3x-2)(-7x)\ dx \\
&= 7\int_1^2 x^3-3x^2+2x\ dx \\
&= -\frac{7}{4}. \quad \square
\end{aligned}$$

Problem 4: Use the change of variables to polar coordinates to show that the area of a circle of radius R is πR^2, i.e., prove that

$$\int\int_{D(0,R)} 1\ dxdy = \pi R^2.$$

Solution: We use the polar coordinates change of variables

$$(x,y) = (r\cos\theta, r\sin\theta) \quad \text{with} \quad (r,\theta) \in \Omega = [0,R]\times[0,2\pi).$$

Recall that $dxdy = rd\theta dr$. Then,

$$\int\int_{D(0,R)} 1\ dxdy = \int_0^R\int_0^{2\pi} r\ d\theta dr = 2\pi\int_0^R rdr = \pi R^2,$$

which is equal to the area of a circle of radius R. $\square$

Problem 5: Compute the area of the ellipse with semi axes a and b, i.e., compute

$$\int\int_D 1\ dx\ dy,$$

where

$$D = \left\{(x,y)\mid \frac{x^2}{a^2}+\frac{y^2}{b^2}\le 1\right\},$$

with $a, b > 0$.

Solution: We use the following change of variables:

$$(x, y) = (ar\cos\theta, br\sin\theta) \quad \text{with} \quad (r, \theta) \in \Omega = [0,1] \times [0, 2\pi).$$

The Jacobian of the change of variables $(x, y) \in D \longrightarrow (r, \theta) \in \Omega$ is

$$\begin{aligned} dxdy &= \left|\frac{\partial x}{\partial r} \cdot \frac{\partial y}{\partial \theta} - \frac{\partial x}{\partial \theta} \cdot \frac{\partial y}{\partial r}\right| d\theta dr \\ &= \left|a\cos\theta \cdot br\cos\theta - (-ar\sin\theta) \cdot b\sin\theta\right| d\theta dr \\ &= abr\left|\cos^2\theta + \sin^2\theta\right| d\theta dr \\ &= abr\ d\theta dr, \end{aligned}$$

since $\cos^2\theta + \sin^2\theta = 1$ for any $\theta \in [0, 2\pi)$. Then,

$$\int\int_D 1\ dxdy = \int_0^1 \int_0^{2\pi} abr\ d\theta dr = ab \cdot 2\pi \int_0^1 rdr = \pi ab,$$

which is equal to the area of a ellipse with semi axes a and b. □

Problem 6: Which number is larger, e^π or π^e?

Solution: We will show that $\pi^e < e^\pi$.

By taking the natural logarithm, it is easy to see that

$$\pi^e < e^\pi \iff e\ln(\pi) < \pi \iff \frac{\ln(\pi)}{\pi} < \frac{1}{e} = \frac{\ln(e)}{e}. \quad (8.13)$$

Let $f : (0, \infty) \to \mathbb{R}$ given by

$$f(x) = \frac{\ln(x)}{x}.$$

Then,

$$f'(x) = \frac{1 - \ln(x)}{x^2}.$$

The function $f(x)$ has one critical point corresponding to $x = e$, is increasing on the interval $(0, e)$ and is decreasing on the interval (e, ∞).

We conclude that $f(x)$ has a global maximum point at $x = e$, i.e., $f(x) < f(e) = \frac{1}{e}$ for all $x > 0$ with $x \neq e$, and therefore

$$f(\pi) = \frac{\ln(\pi)}{\pi} < \frac{1}{e},$$

which is equivalent to $\pi^e < e^\pi$; cf. (8.13). □

Problem 7: Show that

$$\frac{e^x + e^y}{2} \geq e^{\frac{x+y}{2}}, \quad \forall\, x, y \in \mathbb{R}. \tag{8.14}$$

Solution: Let $e^x = a$ and $e^y = b$. Note that $a, b > 0$, and that

$$e^{\frac{x+y}{2}} \;=\; \sqrt{e^{x+y}} \;=\; \sqrt{e^x \cdot e^y} \;=\; \sqrt{ab}.$$

Then, (8.14) can be written as

$$\begin{aligned} \frac{a+b}{2} \geq \sqrt{ab} &\iff a + b \;-\; 2\sqrt{ab} \;\geq\; 0 \\ &\iff \left(\sqrt{a} - \sqrt{b}\right)^2 \;\geq\; 0, \end{aligned}$$

which is what we wanted to show. □

Problem 8: Let $u, v : [0, \infty) \to [0, \infty)$ be two continuous functions with positive values. Assume that there exists a constant $M > 0$ such that

$$u(x) \;\leq\; M + \int_0^x u(t)v(t)\,dt, \quad \forall\, x \geq 0. \tag{8.15}$$

Show that

$$u(x) \;\leq\; M \exp\left(\int_0^x v(t)\,dt\right), \quad \forall\, x \geq 0. \tag{8.16}$$

Solution: Define the function $w : [0, \infty) \to [0, \infty)$ as

$$w(x) \;=\; \left(M + \int_0^x u(t)v(t)\,dt\right)\, \exp\left(-\int_0^x v(t)\,dt\right). \tag{8.17}$$

Recall that

$$\left(\int_0^x u(t)v(t)\,dt\right)' = u(x)v(x); \quad \left(\int_0^x v(t)\,dt\right)' = v(x),$$

where the derivative is computed with respect to x.

Using the product rule, we find that

$$\begin{aligned} w'(x) \;&=\; u(x)v(x)\,\exp\left(-\int_0^x v(t)\,dt\right) \\ &\quad+\; \left(M + \int_0^x u(t)v(t)\,dt\right)\left(-v(x)\,\exp\left(-\int_0^x v(t)\,dt\right)\right) \\ &=\; \left(u(x) - M - \int_0^x u(t)v(t)\,dt\right)\; v(x)\exp\left(-\int_0^x v(t)dt\right) \end{aligned} \tag{8.18}$$

Using (8.15) and the fact that $v(x) \geq 0$ for all $x \geq 0$, we conclude from (8.18) that

$$w'(x) \ \leq \ 0, \ \ \forall \, x \geq 0.$$

In other words, $w(x)$ is a decreasing function on the interval $[0, \infty)$ and therefore $w(0) \geq w(x)$ for all $x \geq 0$. Since $w(0) = M$, and using (8.15), it follows that

$$\begin{aligned} M \ &\geq \ \left(M + \int_0^x u(t)v(t) \, dt \right) \exp\left(- \int_0^x v(t) \, dt \right) \\ &\geq \ u(x) \exp\left(- \int_0^x v(t) \, dt \right) \quad \forall \, x \geq 0, \end{aligned}$$

which is equivalent to (8.16). $\square$

Problem 9: Let $V(S,t) = \exp(-ax - b\tau)u(x,\tau)$, where

$$x = \ln\left(\frac{S}{K}\right), \quad \tau = \frac{(T-t)\sigma^2}{2}, \quad a = \frac{r-q}{\sigma^2} - \frac{1}{2}, \quad b = \left(\frac{r-q}{\sigma^2} + \frac{1}{2}\right)^2 + \frac{2q}{\sigma^2}.$$

This is the change of variables that reduces the Black–Scholes PDE for $V(S,t)$ to the heat equation for $u(x,\tau)$.

(i) Show that the boundary condition $V(S,T) = \max(S-K,0)$ for the European call option becomes the following boundary condition for $u(x,\tau)$ at time $\tau = 0$:

$$u(x,0) \ = \ K \exp(ax) \ \max(e^x - 1, 0).$$

(ii) Show that the boundary condition $V(S,T) = \max(K-S,0)$ for the European put option becomes

$$u(x,0) \ = \ K \exp(ax) \ \max(1 - e^x, 0).$$

Solution: If $t = T$, then $\tau = 0$, and therefore

$$V(S,T) \ = \ \exp(-ax)u(x,0). \tag{8.19}$$

Here, $x = \ln\left(\frac{S}{K}\right)$, which can also be written as $S = Ke^x$.

(i) For a call option, $V(S,T) = \max(S-K,0)$. From (8.19), we find that

$$\begin{aligned} u(x,0) \ &= \ \exp(ax)V(S,T) \ = \ \exp(ax)\max(S-K,0) \\ &= \ \exp(ax)\max(Ke^x - K, 0) \\ &= \ K\exp(ax)\max(e^x - 1, 0). \end{aligned}$$

(ii) For a put option, $V(S,T) = \max(K-S,0)$. From (8.19), we find that

$$\begin{aligned} u(x,0) &= \exp(ax)V(S,T) \;=\; \exp(ax)\max(K-S,0) \\ &= \exp(ax)\max(K-Ke^x,0) \\ &= K\exp(ax)\max(1-e^x,0). \quad \square \end{aligned}$$

Problem 10: Solve for a and b the following system of equations:

$$\begin{cases} 2a+1-\frac{2(r-q)}{\sigma^2} = 0; \\ b+a^2+a\left(1-\frac{2(r-q)}{\sigma^2}\right)-\frac{2r}{\sigma^2} = 0. \end{cases}$$

Solution: From the first equation, it is easy to see that

$$a = \frac{r-q}{\sigma^2}-\frac{1}{2}. \tag{8.20}$$

Using (8.20), we note that the second equation can be written as

$$\begin{aligned} b &= -a^2 - a\left(1-\frac{2(r-q)}{\sigma^2}\right)+\frac{2r}{\sigma^2} \\ &= -\left(\frac{r-q}{\sigma^2}-\frac{1}{2}\right)^2 - \left(\frac{r-q}{\sigma^2}-\frac{1}{2}\right)\cdot 2\left(\frac{1}{2}-\frac{r-q}{\sigma^2}\right) + \frac{2r}{\sigma^2} \\ &= -\left(\frac{r-q}{\sigma^2}-\frac{1}{2}\right)^2 + 2\left(\frac{r-q}{\sigma^2}-\frac{1}{2}\right)^2 + \frac{2r}{\sigma^2} \\ &= \left(\frac{r-q}{\sigma^2}-\frac{1}{2}\right)^2+\frac{2r}{\sigma^2} \\ &= \left(\frac{r-q}{\sigma^2}+\frac{1}{2}\right)^2-\frac{2(r-q)}{\sigma^2}+\frac{2r}{\sigma^2} \\ &= \left(\frac{r-q}{\sigma^2}+\frac{1}{2}\right)^2+\frac{2q}{\sigma^2}. \quad \square \end{aligned}$$

Problem 11: What does the boundary condition $V(B,t) = R$ for a down–and–out call with barrier B and rebate $R > 0$ correspond to for the function $u(x,\tau)$ given by

$$V(S,t) = \exp(-ax-b\tau)u(x,\tau), \tag{8.21}$$

where

$$x = \ln\left(\frac{S}{K}\right); \quad \tau = \frac{(T-t)\sigma^2}{2};$$

$$a = \frac{r-q}{\sigma^2} - \frac{1}{2}; \quad b = \left(\frac{r-q}{\sigma^2} + \frac{1}{2}\right)^2 + \frac{2q}{\sigma^2}.$$

Solution: Note that $S = B$ corresponds to $x = \ln\left(\frac{B}{K}\right)$ and $0 \leq t \leq T$ corresponds to $0 \leq \tau \leq \frac{T\sigma^2}{2}$. From (8.21), it follows that

$$u(x, \tau) = \exp(ax + b\tau)V(S, t).$$

The boundary condition corresponding to $V(B, t) = R$ for all $0 \leq t \leq T$ is

$$\begin{aligned} u(x, \tau) &= \exp\left(a \ln\left(\frac{B}{K}\right) + b\tau\right) V(B, t) \\ &= \left(\frac{B}{K}\right)^a e^{b\tau} V(B, t) \\ &= \left(\frac{B}{K}\right)^a e^{b\tau} R, \quad \forall\, 0 \leq \tau \leq \frac{T\sigma^2}{2}. \quad \square \end{aligned}$$

Problem 12: Assume that the function $V(S, I, t)$ satisfies the PDE

$$\frac{\partial V}{\partial t} + \ln S \frac{\partial V}{\partial I} + \frac{1}{2}\sigma^2 S^2 \frac{\partial^2 V}{\partial S^2} + rS\frac{\partial V}{\partial S} - rV = 0. \tag{8.22}$$

Consider the following change of variables:

$$V(S, I, t) = F(y, t),$$

where

$$y = \frac{I + (T-t)\ln S}{T}.$$

Show that $F(y, t)$ satisfies the following PDE:

$$\frac{\partial F}{\partial t} + \frac{\sigma^2 (T-t)^2}{2T^2}\frac{\partial^2 F}{\partial y^2} + \left(r - \frac{\sigma^2}{2}\right)\frac{T-t}{T}\frac{\partial F}{\partial y} - rF = 0.$$

Solution: Let $V(S, I, t) = F(y, t)$ with

$$y = \frac{I + (T-t)\ln S}{T}.$$

Note that

$$\frac{\partial y}{\partial t} = -\frac{\ln S}{T}; \quad \frac{\partial y}{\partial I} = \frac{1}{T}; \quad \frac{\partial y}{\partial S} = \frac{1}{S}\frac{T-t}{T}.$$

Using Chain Rule, it is easy to see that

$$\begin{aligned}
\frac{\partial V}{\partial t} &= \frac{\partial F}{\partial t} + \frac{\partial F}{\partial y}\frac{\partial y}{\partial t} = \frac{\partial F}{\partial t} - \frac{\ln S}{T}\frac{\partial F}{\partial y}; \\
\frac{\partial V}{\partial I} &= \frac{\partial F}{\partial y}\frac{\partial y}{\partial I} = \frac{1}{T}\frac{\partial F}{\partial y}; \\
\frac{\partial V}{\partial S} &= \frac{\partial F}{\partial y}\frac{\partial y}{\partial S} = \frac{1}{S}\frac{T-t}{T}\frac{\partial F}{\partial y}; \\
\frac{\partial^2 V}{\partial S^2} &= -\frac{1}{S^2}\frac{T-t}{T}\frac{\partial F}{\partial y} + \frac{1}{S}\frac{T-t}{T}\frac{\partial^2 F}{\partial y^2}\frac{\partial y}{\partial S} \\
&= -\frac{1}{S^2}\frac{T-t}{T}\frac{\partial F}{\partial y} + \frac{1}{S^2}\frac{(T-t)^2}{T^2}\frac{\partial^2 F}{\partial y^2} \\
&= \frac{1}{S^2}\left(\frac{(T-t)^2}{T^2}\frac{\partial^2 F}{\partial y^2} - \frac{T-t}{T}\frac{\partial F}{\partial y}\right).
\end{aligned}$$

Then, the PDE (8.22) for $V(S,I,t)$ becomes the following PDE for $F(y,t)$:

$$\begin{aligned}
0 &= \frac{\partial V}{\partial t} + \ln S\frac{\partial V}{\partial I} + \frac{1}{2}\sigma^2 S^2\frac{\partial^2 V}{\partial S^2} + rS\frac{\partial V}{\partial S} - rV \\
&= \frac{\partial F}{\partial t} - \frac{\ln S}{T}\frac{\partial F}{\partial y} + \ln S\frac{1}{T}\frac{\partial F}{\partial y} \\
&\quad + \frac{1}{2}\sigma^2\left(\frac{(T-t)^2}{T^2}\frac{\partial^2 F}{\partial y^2} - \frac{T-t}{T}\frac{\partial F}{\partial y}\right) + r\frac{T-t}{T}\frac{\partial F}{\partial y} - rF \\
&= \frac{\partial F}{\partial t} + \frac{\sigma^2(T-t)^2}{2T^2}\frac{\partial^2 F}{\partial y^2} + \left(r - \frac{\sigma^2}{2}\right)\frac{T-t}{T}\frac{\partial F}{\partial y} - rF. \quad \square
\end{aligned}$$

Problem 13: Assume that the function $V(S,I,t)$ satisfies the following PDE:

$$\frac{\partial V}{\partial t} + S\frac{\partial V}{\partial I} + \frac{1}{2}\sigma^2 S^2\frac{\partial^2 V}{\partial S^2} + rS\frac{\partial V}{\partial S} - rV = 0. \tag{8.23}$$

Consider the following change of variables:

$$V(S,I,t) = S\,H(R,t), \quad \text{where} \quad R = \frac{I}{S}. \tag{8.24}$$

Show that $H(R,t)$ satisfies the following PDE:

$$\frac{\partial H}{\partial t} + \frac{1}{2}\sigma^2 R^2\frac{\partial^2 H}{\partial R^2} + (1 - rR)\frac{\partial H}{\partial R} = 0. \tag{8.25}$$

Solution: Let $V(S,I,t) = S\ H(R,t)$, with

$$R = \frac{I}{S}.$$

Using Chain Rule, we find that

$$\begin{aligned}
\frac{\partial V}{\partial t} &= S\,\frac{\partial H}{\partial t};\\
\frac{\partial V}{\partial I} &= S\,\frac{\partial H}{\partial R}\,\frac{\partial R}{\partial I} = \frac{\partial H}{\partial R};\\
\frac{\partial V}{\partial S} &= H + S\,\frac{\partial H}{\partial R}\,\frac{\partial R}{\partial S}\\
&= H + S\,\frac{\partial H}{\partial R}\left(-\frac{I}{S^2}\right)\\
&= H - \frac{I}{S}\,\frac{\partial H}{\partial R}\\
&= H - R\,\frac{\partial H}{\partial R};\\
\frac{\partial^2 V}{\partial S^2} &= \frac{\partial H}{\partial R}\,\frac{\partial R}{\partial S} - \frac{\partial R}{\partial S}\,\frac{\partial H}{\partial R} - R\,\frac{\partial^2 H}{\partial R^2}\,\frac{\partial R}{\partial S} = R\,\frac{I}{S^2}\,\frac{\partial^2 H}{\partial R^2}\\
&= \frac{R^2}{S}\,\frac{\partial^2 H}{\partial R^2}.
\end{aligned}$$

By substituting into (8.23), it follows that

$$\begin{aligned}
0 &= \frac{\partial V}{\partial t} + S\frac{\partial V}{\partial I} + \frac{1}{2}\sigma^2 S^2\frac{\partial^2 V}{\partial S^2} + rS\frac{\partial V}{\partial S} - rV\\
&= S\frac{\partial H}{\partial t} + S\frac{\partial H}{\partial R} + \frac{1}{2}\sigma^2 S^2\frac{R^2}{S}\,\frac{\partial^2 H}{\partial R^2} + rS\left(H - R\,\frac{\partial H}{\partial R}\right) - rSH\\
&= S\frac{\partial H}{\partial t} + \frac{1}{2}\sigma^2 SR^2\,\frac{\partial^2 H}{\partial R^2} + S(1 - rR)\,\frac{\partial H}{\partial R}.
\end{aligned}$$

By dividing by S, we conclude that $H(R,t)$ satisfies the PDE

$$\frac{\partial H}{\partial t} + \frac{1}{2}\sigma^2 R^2\frac{\partial^2 H}{\partial R^2} + (1 - rR)\,\frac{\partial H}{\partial R} = 0,$$

which is the same as (8.25). □

Problem 14: The price of a non-dividend-paying asset is lognormally distributed. Assume that the spot price is 40, the volatility is 30%, and the interest rates are constant at 5%. Fill in the Black–Scholes values of the ITM put options on the asset in the table below:

Option Type	Strike	Maturity	Value
Put	45	6 months	
Put	45	3 months	
Put	48	6 months	
Put	48	3 months	
Put	51	6 months	
Put	51	3 months	

For which of these options is the intrinsic value $\max(K - S, 0)$ larger than the Black–Scholes value of the option (in which case the corresponding American put is guaranteed to be worth more than the European put)?

Solution: The answers are summarized in the table below:

Option Type	Strike	Maturity	Value	$K - S$
Put	45	6 months	5.8196	5
Put	45	3 months	5.3403	5
Put	48	6 months	8.0325	8
Put	48	3 months	7.8234	8
Put	51	6 months	10.4862	11
Put	51	3 months	10.5476	11

In general, the values of deep–in–the money European put options are lower than the premium $K - S$. This feature was observed for the options priced here: the values of the 51–puts and of the three months 48–put are below their intrinsic value $K - S$.

Also, note that for the 51–puts (i.e., for deep in the money puts), the values of the short dated options are higher than the values of the long dated options. This is to be expected; for example, if the spot price is 0, the value of a European put is Ke^{-rT}, in which case longer dated puts are worth less than short dated ones. □

Problem 15: Show that the premium of the Black–Scholes value of a European call option over its intrinsic value $\max(S - K, 0)$ is largest at the money. In other words, show that the maximum value of

$$C_{BS}(S) - \max(S - K, 0)$$

is obtained for $S = K$, where $C_{BS}(S)$ is the Black–Scholes value of the plain vanilla European call option with strike K and spot price S.

Solution: Let $f(S) = C_{BS}(S) - \max(S - K, 0)$. It is easy to see that

$$f(S) = \begin{cases} C_{BS}(S), & \text{if } S \leq K; \\ C_{BS}(S) - S + K, & \text{if } S > K. \end{cases}$$

Note that $f(S)$ is a continuous function, but it is not differentiable at $S = K$.

For $S < K$, the function $f(S)$ is the value of a call with strike K, and therefore is increasing.

For $S > K$, we find that

$$f'(S) = \Delta(C_{BS}) - 1 = e^{-qT}N(d_1) - 1 < N(d_1) - 1 = -N(-d_1) < 0,$$

and therefore the function $f(S)$ is decreasing.

We conclude that $f(S)$ has an absolute maximum point at $S = K$. □

Problem 16: Use the formula

$$V(S, K, t) = C(S, K, t) - \left(\frac{B}{S}\right)^{2a} C\left(\frac{B^2}{S}, K, t\right), \tag{8.26}$$

where $a = \frac{r-q}{\sigma^2} - \frac{1}{2}$, to find the value of a six months down–and–out call on a non–dividend–paying asset with price following a lognormal distribution with 30% volatility and spot price 40. The barrier is $B = 35$ and the strike for the call is $K = 40$. The risk–free interest rate is constant at 5%.

Solution: The value of the down–and–out call is \$3.398883. □

Problem 17: Show that the value of a down–and–out call with barrier B less than the strike K of the call, i.e., $B < K$, converges to the value of a plain vanilla call with strike K when $B \searrow 0$. For simplicity, assume that the underlying asset does not pay dividends and that interest rates are zero.

Solution: Let $t = 0$ in formula (8.26). For $r = q = 0$, we find that $a = -\frac{1}{2}$. Therefore, the value of the down–and–out call is

$$\begin{aligned} V(S) &= C(S) - \frac{S}{B} C\left(\frac{B^2}{S}\right) \\ &= C(S) - \frac{S}{B}\left(\frac{B^2}{S}N(d_1) - KN(d_2)\right) \\ &= C(S) - BN(d_1) + \frac{SK}{B}N(d_2), \end{aligned} \tag{8.27}$$

where $C(S)$ is the value of the plain vanilla call with strike K and

$$d_1 = \frac{\ln\left(\frac{B^2}{SK}\right) + \frac{\sigma^2}{2}T}{\sigma\sqrt{T}} \quad \text{and} \quad d_2 = \frac{\ln\left(\frac{B^2}{SK}\right) - \frac{\sigma^2}{2}T}{\sigma\sqrt{T}}.$$

Note that $0 < N(d_1) < 1$. Then,

$$\lim_{B\searrow 0} BN(d_1) = 0. \tag{8.28}$$

Using l'Hôpital's rule, we obtain that

$$\begin{aligned}\lim_{B\searrow 0}\frac{SK}{B}\,N(d_2) &= SK\lim_{B\searrow 0}\frac{N(d_2)}{B} = SK\lim_{B\searrow 0}N'(d_2)\frac{\partial d_2}{\partial B}\\ &= SK\lim_{B\searrow 0}\frac{1}{\sqrt{2\pi}}\exp\left(-\frac{d_2^2}{2}\right)\cdot\frac{2}{B\sigma\sqrt{T}}\\ &= \frac{2SK}{\sigma\sqrt{2\pi T}}\lim_{B\searrow 0}\exp\left(-\frac{d_2^2}{2}-\ln(B)\right). \qquad (8.29)\end{aligned}$$

As $B\searrow 0$, the term $-\frac{d_2^2}{2}-\ln(B)$ is on the order of $-(\ln(B))^2$, and therefore

$$\lim_{B\searrow 0}\left(-\frac{d_2^2}{2}-\ln(B)\right) = -\infty. \qquad (8.30)$$

From (8.29) and (8.30) we find that

$$\lim_{B\searrow 0}\frac{SK}{B}\,N(d_2) = 0, \qquad (8.31)$$

and, from (8.27), (8.28), and (8.31), we conclude that

$$\lim_{B\searrow 0}V(S) = \lim_{B\searrow 0}\left(C(S) - BN(d_1) + \frac{SK}{B}N(d_2)\right) = C(S). \quad \square$$

Problem 18: Compute the Delta and Gamma of a down–and–out call with $B < K$.

Solution: We rewrite formula (8.26) to emphasize the dependence of the value $V(S)$ of the down–and–out call on the spot price S of the underlying asset as follows:

$$V(S) = C_{BS}(S) - \frac{B^{2a}}{S^{2a}}\,C_{BS}\left(\frac{B^2}{S}\right). \qquad (8.32)$$

By differentiating (8.32) with respect to S, we obtain that

$$\Delta(V) = \Delta(C_{BS})(S) + \frac{2aB^{2a}}{S^{2a+1}}\,C_{BS}\left(\frac{B^2}{S}\right) + \frac{B^{2a+2}}{S^{2a+2}}\,\Delta(C_{BS})\left(\frac{B^2}{S}\right),$$

where $\Delta(C_{BS})(x) = e^{-qT}N(d_1(x))$, with

$$d_1(x) = \frac{\ln\left(\frac{x}{K}\right) + \left(r-q+\frac{\sigma^2}{2}\right)T}{\sigma\sqrt{T}}. \qquad (8.33)$$

Similarly,

$$\Gamma(V) = \Gamma(C_{BS})(S) - \frac{2a(2a+1)B^{2a}}{S^{2a+2}} C_{BS}\left(\frac{B^2}{S}\right) - \frac{(4a+2)B^{2a+2}}{S^{2a+3}} \Delta(C_{BS})\left(\frac{B^2}{S}\right) - \frac{B^{2a+4}}{S^{2a+4}} \Gamma(C_{BS})\left(\frac{B^2}{S}\right),$$

where

$$\Gamma(C_{BS})(x) = \frac{e^{-qT}}{x\sigma\sqrt{2\pi T}} \exp\left(-\frac{d_1(x)^2}{2}\right),$$

with $d_1(x)$ given by (8.33). □

Problem 19: (i) Show that the value of a plain vanilla European call option on a non–dividend–paying asset is always greater than the intrinsic premium $\max(S-K,0)$.

(ii) Conclude that plain vanilla American call options on non–dividend–paying assets are never optimal to exercise early.

Solution: (i) If $S < K$, the intrinsic premium of the call option is 0, since $\max(S-K,0) = 0$ for $S < K$, and therefore the value $C > 0$ of the call option is greater than the zero intrinsic premium of the option.

If $S \geq K$, recall from the Put–Call parity that the value C of a plain vanilla European call option on a non–dividend–paying asset can be written as

$$C = P + S - Ke^{-rT}.$$

Since $P > 0$, it follows that

$$C > S - Ke^{-rT} > S - K.$$

Therefore $C > \max(S-K,0) = S-K$, for any $S \geq K$, and we conclude that the value of a plain vanilla European call option on a non–dividend–paying asset is always greater than the intrinsic premium of the option.

(ii) Let $V_{Amer}(S,t)$ and $V_{Eur}(S,t)$ be the values of an American call and of a European call option, respectively, on a non–dividend–paying asset, and assume that the American option is exercised at time t. Then, $V_{Amer}(S,t) = S-K$; note that the spot price S must be greater than the strike K for early exercise to make sense.

However, we established in part (i) that the value of a plain vanilla European call option on a non–dividend–paying asset is always greater than the intrinsic premium $\max(S-K,0)$, i.e., $V_{Eur}(S,t) > S-K$. This results in a contradiction: the American option whose value $V_{Amer}(S,t)$ must be at least as much as the value $V_{Eur}(S,t)$ of the corresponding European option, was exercised for a value $S - K$ which is smaller than the value of the European option. □

Problem 20: Show that the Black–Scholes value of a European call option on an asset paying dividends continuously at the rate $q > 0$ is smaller than the intrinsic value $S - K$ is the spot price S is large enough, regardless of how small q is.

Solution: We want to show that, if the dividend rate of the underlying asset is $q > 0$, then $C_{BS}(S,K) < S - K$ for S large enough.

Note that

$$C_{BS}(S,K) \;=\; Se^{-qT}N(d_1) - Ke^{-rT}N(d_2) \;<\; Se^{-qT},$$

since $N(d_1) < 1$ and $N(d_2) > 0$.

If $Se^{-qT} < S - K$, which is equivalent to $S > \frac{K}{1-e^{-qT}} > 0$ since $q > 0$, it follows that $C_{BS}(S,K) < S - K$. We conclude that, if $S > \frac{K}{1-e^{-qT}}$, then

$$C_{BS}(S,K) < S - K,$$

which is what we wanted to show. □

Problem 21: For the same maturity, options with different strikes trade simultaneously. The goal of this problem is to compute the rate of change of the implied volatility as a function of the strikes of the options.

In other words, assume that S, T, q and r are given, and let $C(K)$ be the (known) value of a call option with maturity T and strike K. Assume that options with all strikes K exist. Define the implied volatility $\sigma_{imp}(K)$ as the unique solution to

$$C(K) \;=\; C_{BS}(K, \sigma_{imp}(K)),$$

where $C_{BS}(K, \sigma_{imp}(K)) \;=\; C_{BS}(S, K, T, \sigma_{imp}(K), r, q)$ denotes the Black–Scholes value of a call option with strike K on an underlying asset following a lognormal model with volatility $\sigma_{imp}(K)$. Find an implicit differential equation satisfied by $\sigma_{imp}(K)$, i.e., find

$$\frac{\partial \sigma_{imp}(K)}{\partial K}$$

as a function of $\sigma_{imp}(K)$.

Solution: We first find the partial derivative of the Black–Scholes value $C_{BS}(K)$ of a call option with respect to its strike K. Recall that

$$C_{BS}(S, K) = Se^{-qT}N(d_1) - Ke^{-rT}N(d_2).$$

Then,

$$\frac{\partial C_{BS}}{\partial K} = Se^{-qT}N'(d_1)\,\frac{\partial d_1}{\partial K} - e^{-rT}N(d_2) - Ke^{-rT}N'(d_2)\,\frac{\partial d_2}{\partial K}. \quad (8.34)$$

Also, recall that

$$Se^{-qT}N'(d_1) = Ke^{-rT}N'(d_2); \quad (8.35)$$

cf. Lemma 3.15 of [1]. From (8.34) and (8.35), we find that

$$\frac{\partial C_{BS}}{\partial K} = -e^{-rT}N(d_2) + Ke^{-rT}N'(d_2)\left(\frac{\partial d_1}{\partial K} - \frac{\partial d_2}{\partial K}\right). \quad (8.36)$$

Since $d_1 = d_2 + \sigma\sqrt{T}$, it follows that

$$\frac{\partial d_1}{\partial K} = \frac{\partial(d_2 + \sigma\sqrt{T})}{\partial K} = \frac{\partial d_2}{\partial K}.$$

We conclude from (8.36) that

$$\frac{\partial C_{BS}}{\partial K} = -e^{-rT}N(d_2). \quad (8.37)$$

We now differentiate formula

$$C(K) = C_{BS}(K, \sigma_{imp}(K)),$$

the definition of $\sigma_{imp}(K)$, with respect to K. Note that $C(K)$ is assumed to be derived from market prices, and known for all the values K. Using Chain Rule and (8.37), we find that

$$\begin{aligned} \frac{\partial C}{\partial K} &= \frac{\partial C_{BS}}{\partial K} + \frac{\partial C_{BS}}{\partial \sigma} \cdot \frac{\partial \sigma_{imp}(K)}{\partial K} \\ &= -e^{-rT}N(d_2) + \text{vega}(C_{BS})\frac{\partial \sigma_{imp}(K)}{\partial K}, \end{aligned} \quad (8.38)$$

where

$$\begin{aligned} \text{vega}(C_{BS}) &= \frac{\partial C_{BS}}{\partial \sigma} = Se^{-qT}\sqrt{\frac{T}{2\pi}}e^{-\frac{d_1^2}{2}} = \sqrt{T}Se^{-qT}N'(d_1) \\ &= \sqrt{T}Ke^{-rT}N'(d_2) = Ke^{-rT}\sqrt{\frac{T}{2\pi}}e^{-\frac{d_2^2}{2}}; \end{aligned}$$

cf. (8.35). We conclude that the implied differential equation (8.38) can be written as

$$K e^{-rT} \sqrt{\frac{T}{2\pi}} e^{-\frac{d_2^2}{2}} \frac{\partial \sigma_{imp}(K)}{\partial K} = \frac{\partial C}{\partial K} + e^{-rT} N(d_2),$$

with

$$d_2 = \frac{\ln\left(\frac{S}{K}\right) + (r-q)T}{\sigma_{imp}(K)\sqrt{T}} - \frac{\sigma_{imp}(K)\sqrt{T}}{2}. \quad \square$$

Chapter 9

Lagrange multipliers. Portfolio optimization.

9.1 Exercises

1. Find the maximum and minimum of the function

$$f(x_1, x_2, x_3, x_4) \;=\; x_1 - 2x_2 + 3x_3 - 4x_4,$$

subject to the constraints $x_1^2 + x_2^2 - x_4^2 = 1$ and $x_1^2 + 2x_2^2 + 3x_3^2 = 6$.

2. Prove the arithmetic–geometric–harmonic means inequality:

$$\frac{1}{n}\sum_{i=1}^{n} x_i \;\geq\; \left(\prod_{i=1}^{n} x_i\right)^{\frac{1}{n}} \;\geq\; \frac{n}{\sum_{i=1}^{n}\frac{1}{x_i}}, \quad \forall\; x_i > 0,\; i = 1:n. \tag{9.1}$$

Note: The left term of (9.1) is the arithmetic mean of $(x_i)_{i=1:n}$; the middle term of (9.1) is the geometric mean of $(x_i)_{i=1:n}$; the right term of (9.1) is the harmonic mean of $(x_i)_{i=1:n}$.

3. Consider five tradable assets with the following expected values and standard deviations of the rates of return of the assets:

$$\begin{array}{llllll} \mu_1 = 0.08; & \mu_2 = 0.10; & \mu_3 = 0.13; & \mu_4 = 0.15; & \mu_5 = 0.20; \\ \sigma_1 = 0.14; & \sigma_2 = 0.18; & \sigma_3 = 0.23; & \sigma_4 = 0.25; & \sigma_5 = 0.35. \end{array}$$

The correlation matrix of the rates of return is

$$\Omega \;=\; \begin{pmatrix} 1 & -0.3 & 0.4 & 0.25 & -0.2 \\ -0.3 & 1 & -0.1 & -0.2 & 0.15 \\ 0.4 & -0.1 & 1 & 0.35 & 0.25 \\ 0.25 & -0.2 & 0.35 & 1 & -0.15 \\ -0.2 & 0.15 & 0.25 & -0.15 & 1 \end{pmatrix}$$

Assume that it is possible to take both long and short positions of arbitrary size in these assets.

(i) Find the asset allocation for a minimal variance portfolio with 15% expected rate of return and the corresponding minimal standard deviation of the rate of return of the portfolio;

(ii) Find the asset allocation for a maximum expected return portfolio with 25% standard deviation of the rate of return and the corresponding maximal expected rate of return of the portfolio.

9.2 Solutions to Chapter 9 Exercises

Problem 1: Find the maximum and minimum of the function

$$f(x_1, x_2, x_3, x_4) \;=\; x_1 - 2x_2 + 3x_3 - 4x_4,$$

subject to the constraints $x_1^2 + x_2^2 - x_4^2 = 1$ and $x_1^2 + 2x_2^2 + 3x_3^2 = 6$.

Solution: We reformulate the problem as a constrained optimization problem. Let $f : \mathbb{R}^4 \to \mathbb{R}$ and $g : \mathbb{R}^4 \to \mathbb{R}^2$ be defined as follows:

$$\begin{aligned} f(x) &= x_1 - 2x_2 + 3x_3 - 4x_4; \\ g(x) &= \begin{pmatrix} x_1^2 + x_2^2 - x_4^2 - 1 \\ x_1^2 + 2x_2^2 + 3x_3^2 - 6 \end{pmatrix}, \end{aligned}$$

where $x = (x_1, x_2, x_3, x_4)$. We also use the notation $g(x) = \begin{pmatrix} g_1(x) \\ g_2(x) \end{pmatrix}$, with

$$g_1(x) \;=\; x_1^2 + x_2^2 - x_4^2 - 1 \quad \text{and} \quad g_2(x) \;=\; x_1^2 + 2x_2^2 + 3x_3^2 - 6.$$

We want to find the maximum and minimum of $f(x)$ on $\mathbb{R}^4$ subject to the constraint $g(x) = 0$.

We first check that $\operatorname{rank}(\nabla g(x)) = 2$ for any x such that $g(x) = 0$. Note that

$$\nabla g(x) \;=\; \begin{pmatrix} 2x_1 & 2x_2 & 0 & -2x_4 \\ 2x_1 & 4x_2 & 6x_3 & 0 \end{pmatrix}.$$

It is easy to see that $\operatorname{rank}(\nabla g(x)) = 2$, unless

$$0 \;=\; x_1x_2 \;=\; x_1x_3 \;=\; x_1x_4 \;=\; x_2x_3 \;=\; x_2x_4 \;=\; x_3x_4 = 0.$$

This happens if and only if one of the following four conditions is satisfied; in each case, $g(x) \neq 0$, as seen below:

$$\begin{aligned} x_1 = x_2 = x_3 = 0 &\implies g_2(x) = -6 \;\neq\; 0; \\ x_1 = x_2 = x_4 = 0 &\implies g_1(x) = -1 \;\neq\; 0; \\ x_1 = x_3 = x_4 = 0 &\implies g(x) = \begin{pmatrix} x_2^2 - 1 \\ 2x_2^2 - 6 \end{pmatrix} \;\neq\; \begin{pmatrix} 0 \\ 0 \end{pmatrix}; \\ x_2 = x_3 = x_4 = 0 &\implies g(x) = \begin{pmatrix} x_1^2 - 1 \\ x_1^2 - 6 \end{pmatrix} \;\neq\; \begin{pmatrix} 0 \\ 0 \end{pmatrix}. \end{aligned}$$

We conclude that $\operatorname{rank}(\nabla g(x)) = 2$ for any x such that $g(x) = 0$.

The Lagrangian associated to our problem is

$$\begin{aligned} F(x, \lambda) \;=\; & x_1 - 2x_2 + 3x_3 - 4x_4 \;+\; \lambda_1 \left(x_1^2 + x_2^2 - x_4^2 - 1\right) \\ & + \lambda_2 \left(x_1^2 + 2x_2^2 + 3x_3^2 - 6\right). \end{aligned} \tag{9.2}$$

where $\lambda = (\lambda_1, \lambda_2)^t \in \mathbb{R}^2$ is the Lagrange multiplier.

Note that

$$\nabla_{(x,\lambda)} F(x,\lambda) = \begin{pmatrix} 1 + 2\lambda_1 x_1 + 2\lambda_2 x_1 \\ -2 + 2\lambda_1 x_2 + 4\lambda_2 x_2 \\ 3 + 6\lambda_2 x_3 \\ -4 - 2\lambda_1 x_4 \\ x_1^2 + x_2^2 - x_4^2 - 1 \\ x_1^2 + 2x_2^2 + 3x_3^2 - 6 \end{pmatrix}^t.$$

To find the critical points of $F(x,\lambda)$, we must solve $\nabla_{(x,\lambda)} F(x,\lambda) = 0$, which is equivalent to $G(x,\lambda) = 0$, where the function $G : \mathbb{R}^6 \to \mathbb{R}^6$ is given by

$$G(x,\lambda_1,\lambda_2) = \begin{pmatrix} 1 + 2\lambda_1 x_1 + 2\lambda_2 x_1 \\ -2 + 2\lambda_1 x_2 + 4\lambda_2 x_2 \\ 3 + 6\lambda_2 x_3 \\ -4 - 2\lambda_1 x_4 \\ x_1^2 + x_2^2 - x_4^2 - 1 \\ x_1^2 + 2x_2^2 + 3x_3^2 - 6 \end{pmatrix}.$$

The solution to $G(x,\lambda) = 0$ is obtained using a 6–dimensional Newton's method. The gradient of $G(x,\lambda)$ is

$$DG(x,\lambda) = \begin{pmatrix} 2\lambda_1 + 2\lambda_2 & 0 & 0 & 0 & 2x_1 & 2x_1 \\ 0 & 2\lambda_1 + 4\lambda_2 & 0 & 0 & 2x_2 & 4x_2 \\ 0 & 0 & 6\lambda_2 & 0 & 0 & 6x_3 \\ 0 & 0 & 0 & -2\lambda_1 & -2x_4 & 0 \\ 2x_1 & 2x_2 & 0 & -2x_4 & 0 & 0 \\ 2x_1 & 4x_2 & 6x_3 & 0 & 0 & 0 \end{pmatrix}.$$

Let $z = (x,\lambda)$. The Newton's method recursion for solving $G(x,\lambda) = G(z) = 0$ can be written as

$$z_{k+1} = z_k - (DG(z_k))^{-1} G(z_k), \quad \forall\, k \geq 0. \tag{9.3}$$

As expected, the initial guess z_0 makes a big difference both in terms of the convergence of the recursion (9.3), and in terms of the solution to which the recursion (9.3) would converge. For example, if $z_0 = (1,1,1,1,1,1)$, the recursion (9.3) will not converge, and, if $z_0 = (1,1,1,1,0,0)$, the matrix $DG(z_0)$ is singular and Newton's iteration fails after one step.

We find two solutions for $G(z) = 0$:

$$\begin{pmatrix} -2.1377 \\ 0.6834 \\ -0.4067 \\ 2.0091 \\ -0.9954 \\ 1.2293 \end{pmatrix}, \tag{9.4}$$

corresponding to the initial guess $z_0 = (-1, 1, -1, 1, -1, 1)$, and

$$\begin{pmatrix} 2.1377 \\ -0.6834 \\ 0.4067 \\ -2.0091 \\ 0.9954 \\ -1.2293 \end{pmatrix}, \qquad (9.5)$$

corresponding to the initial guess $z_0 = (1, -1, 1, -1, 1, -1)$.

Recall that every solution of $G(z) = 0$ corresponds to a solution of $\nabla_{(x\lambda)} F(x, \lambda) = 0$. We therefore found two critical points of $F(x, \lambda)$, i.e.,

$$x_0 = \begin{pmatrix} -2.1377 \\ 0.6834 \\ -0.4067 \\ 2.0091 \end{pmatrix} \quad \text{and} \quad \lambda_0 = \begin{pmatrix} -0.9954 \\ 1.2293 \end{pmatrix}, \qquad (9.6)$$

corresponding to (9.4), and

$$x_0 = \begin{pmatrix} 2.1377 \\ -0.6834 \\ 0.4067 \\ -2.0091 \end{pmatrix} \quad \text{and} \quad \lambda_0 = \begin{pmatrix} 0.9954 \\ -1.2293 \end{pmatrix}, \qquad (9.7)$$

corresponding to (9.5).

For the first critical point (9.6), we compute the Hessian $D^2F_0(x)$ of $F_0(x) = f(x) + \lambda_0^t g(x)$, i.e., of

$$\begin{aligned} F_0(x) &= x_1 - 2x_2 + 3x_3 - 4x_4 - 0.9954\left(x_1^2 + x_2^2 - x_4^2 - 1\right) \\ &\quad + 1.2293\left(x_1^2 + 2x_2^2 + 3x_3^2 - 6\right) \end{aligned}$$

and obtain

$$D^2F_0(x) = \begin{pmatrix} 0.4678 & 0 & 0 & 0 \\ 0 & 2.9265 & 0 & 0 \\ 0 & 0 & 7.3761 & 0 \\ 0 & 0 & 0 & 1.9909 \end{pmatrix}.$$

Note that $D^2F_0(x)$ is a positive definite matrix for any $x \in \mathbb{R}^4$. Thus, the point

$$\begin{pmatrix} -2.1377 \\ 0.6834 \\ -0.4067 \\ 2.0091 \end{pmatrix}$$

is a minimum point for $f(x)$. The corresponding minimum value of the function $f(x)$ is -12.7613.

Similarly, for the second critical point (9.7), we find that

$$\begin{aligned} F_0(x) &= x_1 - 2x_2 + 3x_3 - 4x_4 \; + \; 0.9954\left(x_1^2 + x_2^2 - x_4^2 - 1\right) \\ &\quad - \; 1.2293\left(x_1^2 + 2x_2^2 + 3x_3^2 - 6\right). \end{aligned}$$

Thus,

$$D^2F_0(x) \;=\; \begin{pmatrix} -0.4678 & 0 & 0 & 0 \\ 0 & -2.9265 & 0 & 0 \\ 0 & 0 & -7.3761 & 0 \\ 0 & 0 & 0 & -1.9909 \end{pmatrix},$$

which is negative definite for any $x \in \mathbb{R}^4$. We conclude that the point

$$\begin{pmatrix} 2.1377 \\ -0.6834 \\ 0.4067 \\ -2.0091 \end{pmatrix}$$

is a maximum point for $f(x)$. The corresponding maximum value of the function $f(x)$ is 12.7613. □

Problem 2: Prove the arithmetic–geometric–harmonic means inequality:

$$\frac{1}{n}\sum_{i=1}^{n} x_i \;\geq\; \left(\prod_{i=1}^{n} x_i\right)^{\frac{1}{n}} \;\geq\; \frac{n}{\sum_{i=1}^{n}\frac{1}{x_i}}, \quad \forall\, x_i > 0,\; i = 1:n. \tag{9.8}$$

Solution: We look at the two parts of inequality (9.8) separately:

$$\frac{1}{n}\sum_{i=1}^{n} x_i \;\geq\; \left(\prod_{i=1}^{n} x_i\right)^{\frac{1}{n}}, \quad \forall\, x_i > 0,\; i = 1:n; \tag{9.9}$$

$$\left(\prod_{i=1}^{n} x_i\right)^{\frac{1}{n}} \;\geq\; \frac{n}{\sum_{i=1}^{n}\frac{1}{x_i}}, \quad \forall\, x_i > 0,\; i = 1:n. \tag{9.10}$$

We first note that the geometric–harmonic means inequality (9.10) can be obtained from the arithmetic–geometric means inequality (9.9) as follows: Substitute

$$x_i = \frac{1}{y_i}, \quad \forall\, i = 1:n,$$

in (9.10). Note that $y_i > 0$, for all $i = 1:n$. Then, (9.10) becomes

$$\left(\prod_{i=1}^{n} \frac{1}{y_i}\right)^{\frac{1}{n}} \;\geq\; \frac{n}{\sum_{i=1}^{n} y_i}, \quad \forall\, y_i > 0,\; i = 1:n;$$

$$\Longleftrightarrow \quad \frac{\sum_{i=1}^n y_i}{n} \geq \frac{1}{\left(\prod_{i=1}^n \frac{1}{y_i}\right)^{\frac{1}{n}}}, \quad \forall\, y_i > 0, \ i = 1:n;$$

$$\Longleftrightarrow \quad \frac{1}{n}\sum_{i=1}^n y_i \geq \left(\prod_{i=1}^n y_i\right)^{\frac{1}{n}}, \quad \forall\, y_i > 0, \ i = 1:n,$$

which is the same as (9.9)

Thus, if we proved the arithmetic–geometric means inequality (9.9), then (9.10) also holds true and the proof of (9.8) would be complete.

Note that inequality (9.8) is scalable in the sense that, if each variable x_i is multiplied by the same constant $c > 0$, the inequality (9.8) remains the same:

$$\frac{1}{n}\sum_{i=1}^n (cx_i) \geq \left(\prod_{i=1}^n (cx_i)\right)^{\frac{1}{n}} \geq \frac{n}{\sum_{i=1}^n \frac{1}{cx_i}}$$

$$\Longleftrightarrow \quad c \cdot \frac{1}{n}\sum_{i=1}^n x_i \geq \left(c^n \cdot \prod_{i=1}^n x_i\right)^{\frac{1}{n}} \geq \frac{n}{\frac{1}{c} \cdot \sum_{i=1}^n \frac{1}{x_i}}$$

$$\Longleftrightarrow \quad c \cdot \frac{1}{n}\sum_{i=1}^n x_i \geq c \cdot \left(\prod_{i=1}^n x_i\right)^{\frac{1}{n}} \geq c \cdot \frac{n}{\sum_{i=1}^n \frac{1}{x_i}}$$

$$\Longleftrightarrow \quad \frac{1}{n}\sum_{i=1}^n x_i \geq \left(\prod_{i=1}^n x_i\right)^{\frac{1}{n}} \geq \frac{n}{\sum_{i=1}^n \frac{1}{x_i}}.$$

We conclude that the arithmetic–geometric means inequality (9.9) is also scalable. We can therefore assume, without any loss of generality, that $\prod_{i=1}^n x_i = 1$, and inequality (9.9) reduces to

$$\sum_{i=1}^n x_i \geq n, \quad \forall\, x_i > 0, \ i = 1:n, \quad \text{with} \quad \prod_{i=1}^n x_i = 1. \tag{9.11}$$

We formulate problem (9.11) as a constrained optimization problem. Let $U = \prod_{i=1}^n (0, \infty)^n$ and let $x = (x_1, x_2, \ldots, x_n) \in U$. The functions $f : U \to \mathbb{R}$ and $g : U \to \mathbb{R}$ are defined as

$$f(x) = \sum_{i=1}^n x_i; \quad g(x) = \prod_{i=1}^n x_i - 1.$$

We want to minimize $f(x)$ over the set U subject to the constraint $g(x) = 0$.

Step 1: Check that $\text{rank}(\nabla g(x)) = 1$ for any x such that $g(x) = 0$.
By direct computation, we find that

$$\nabla g(x) = \left(\prod_{i \neq 1} x_i \quad \prod_{i \neq 2} x_i \quad \ldots \quad \prod_{i \neq n} x_i \right). \tag{9.12}$$

Note that $\nabla g(x) \neq 0$ since $x_i > 0$, $i = 1 : n$, for all $x \in U$. Therefore, $\text{rank}(\nabla g(x)) = 1$ for any $x \in U$.

Step 2: Find (x_0, λ_0) such that $\nabla_{(x,\lambda)} F(x_0, \lambda_0) = 0$.
The Lagrangian associated to this problem is

$$F(x, \lambda) = \sum_{i=1}^{n} x_i + \lambda \left(\prod_{i=1}^{n} x_i - 1 \right), \tag{9.13}$$

where $\lambda \in \mathbb{R}$ is the Lagrange multiplier. Let $x_0 = (x_{0,i})_{i=1:n} \in U$ and $\lambda_0 \in \mathbb{R}$. From (9.13), we find that $\nabla_{(x,\lambda)} F(x_0, \lambda_0) = 0$ can be written as

$$\begin{cases} 1 + \lambda_0 \prod_{i \neq j} x_{0,i} = 0, & \forall\, j = 1 : n; \\ \prod_{i=1}^{n} x_{0,i} = 1, \end{cases}$$

which is the same as

$$\begin{cases} 1 + \frac{\lambda_0}{x_{0,j}} = 0, & \forall\, j = 1 : n; \\ \prod_{i=1}^{n} x_{0,i} = 1. \end{cases}$$

The only solution to this system satisfying $x_{0,i} > 0$ for all $i = 1 : n$ is

$$x_0 = \begin{pmatrix} 1 \\ 1 \\ \vdots \\ 1 \end{pmatrix} \quad \text{and} \quad \lambda_0 = -1.$$

Step 3.1: Compute $q(v) = v^t\, D^2 F_0(x_0)\, v$.
Since $\lambda_0 = -1$, we find that $F_0(x) = f(x) + \lambda_0^t g(x)$ is given by

$$F_0(x) = \sum_{i=1}^{n} x_i - \prod_{i=1}^{n} x_i + 1.$$

Then,

$$D^2 F_0(1, 1, \ldots, 1) = \begin{pmatrix} 0 & -1 & -1 & \ldots & -1 \\ -1 & 0 & -1 & \ldots & -1 \\ \vdots & \ddots & \ddots & \ddots & \vdots \\ -1 & \ldots & \ldots & -1 & 0 \end{pmatrix}.$$

The quadratic form $q(v) = v^t D^2 F_0(1,1,\ldots,1)v$ can be written as

$$\begin{aligned} q(v) &= -\sum_{1\le i\ne j\le n} v_i v_j \;=\; -2\sum_{1\le i<j\le n} v_i v_j \\ &= -2v_1\left(\sum_{i=2}^n v_i\right) - 2\sum_{2\le i<j\le n} v_i v_j. \end{aligned} \tag{9.14}$$

Step 3.2: Compute $q_{red}(v_{red})$.
We first solve formally the equation $\nabla g(1,1,\ldots,1)\ v = 0$, for an arbitrary vector $v = (v_i)_{i=1:n} \in \mathbb{R}^n$. From (9.12), we find that

$$\nabla g(1,1,\ldots,1) \;=\; (1,1,\ldots,1).$$

Then, $\nabla g(1,1,\ldots,1)\ v = \sum_{i=1}^n v_i = 0$, and therefore $v_1 = -\sum_{i=2}^n v_i$.

Let $v_{red} = \begin{pmatrix} v_2 \\ \vdots \\ v_n \end{pmatrix}$. We substitute $(-\sum_{i=2}^n v_i)$ for v_1 in (9.14), to obtain the reduced quadratic form $q_{red}(v_{red})$, i.e.,

$$\begin{aligned} q_{red}(v_{red}) &= 2\left(\sum_{i=2}^n v_i\right)^2 - 2\sum_{2\le i<j\le n} v_i v_j \\ &= 2\left(\sum_{i=2}^n v_i^2 + 2\sum_{1\le i<j\le n} v_i v_j\right) - 2\sum_{2\le i<j\le n} v_i v_j \\ &= 2\sum_{i=2}^n v_i^2 + 2\sum_{2\le i<j\le n} v_i v_j \\ &= \sum_{i=2}^n v_i^2 + \left(\sum_{i=2}^n v_i\right)^2. \end{aligned}$$

Then, $q_{red}(v_{red}) > 0$ for any $v_{red} \ne (0,0,\ldots,0) \in \mathbb{R}^{n-1}$, and therefore q_{red} is a positive definite quadratic form.

Step 4: Since $q_{red}(v_{red})$ is a positive definite quadratic form, we conclude that $x_0 = (1,1,\ldots,1)$ is a constrained minimum point for $f(x) = \sum_{i=1}^n x_i$ subject to the constraint that $\prod_{i=1}^n x_i = 1$. In other words,

$$\sum_{i=1}^n x_i \;\ge\; n, \;\; \forall\ x_i > 0,\ i = 1:n, \quad \text{with} \quad \prod_{i=1}^n x_i = 1.$$

This is the same as (9.11), which is what we wanted to prove. □

Problem 3: Consider five tradable assets with the following expected values and standard deviations of the rates of return of the assets:

$$\begin{array}{llllllllllllll} \mu_1 & = & 0.08; & \mu_2 & = & 0.10; & \mu_3 & = & 0.13; & \mu_4 & = & 0.15; & \mu_5 & = & 0.20; \\ \sigma_1 & = & 0.14; & \sigma_2 & = & 0.18; & \sigma_3 & = & 0.23; & \sigma_4 & = & 0.25; & \sigma_5 & = & 0.35. \end{array}$$

The correlation matrix of the rates of return is

$$\Omega \;=\; \begin{pmatrix} 1 & -0.3 & 0.4 & 0.25 & -0.2 \\ -0.3 & 1 & -0.1 & -0.2 & 0.15 \\ 0.4 & -0.1 & 1 & 0.35 & 0.25 \\ 0.25 & -0.2 & 0.35 & 1 & -0.15 \\ -0.2 & 0.15 & 0.25 & -0.15 & 1 \end{pmatrix}$$

Assume that it is possible to take both long and short positions of arbitrary size in these assets.

(i) Find the asset allocation for a minimal variance portfolio with 15% expected rate of return and the corresponding minimal standard deviation of the rate of return of the portfolio;

(ii) Find the asset allocation for a maximum expected return portfolio with 25% standard deviation of the rate of return and the corresponding maximal expected rate of return of the portfolio.

Solution: Let

$$\mu \;=\; \begin{pmatrix} \mu_1 \\ \mu_2 \\ \mu_3 \\ \mu_4 \\ \mu_5 \end{pmatrix} \;=\; \begin{pmatrix} 0.08 \\ 0.10 \\ 0.13 \\ 0.15 \\ 0.20 \end{pmatrix}; \qquad (9.15)$$

$$\sigma \;=\; \begin{pmatrix} \sigma_1 \\ \sigma_2 \\ \sigma_3 \\ \sigma_4 \\ \sigma_5 \end{pmatrix} \;=\; \begin{pmatrix} 0.14 \\ 0.18 \\ 0.23 \\ 0.25 \\ 0.35 \end{pmatrix} \qquad (9.16)$$

be the vectors of expected rates of return and of standard deviations of the rates of return of the given assets, respectively.

The covariance matrix M of the rates of return of the assets is given by[1] $M(i,j) = \sigma(i)\sigma(j)\rho_{i,j}$, for all $i,j = 1:5$, where $\rho_{i,j} = \Omega(i,j)$ is the

[1]Using matrix notation,

$$M \;=\; \mathrm{diag}(\sigma)\ \Omega\ \mathrm{diag}(\sigma),$$

where $\mathrm{diag}(\sigma)$ is the diagonal matrix with diagonal entries equal to the standard deviations of the rates of return of the assets.

correlation of the rates of return of assets i and j. Then,

$$M = \begin{pmatrix} 0.0196 & -0.0076 & 0.0129 & 0.0088 & -0.0098 \\ -0.0076 & 0.0324 & -0.0041 & -0.0090 & 0.0095 \\ 0.0129 & -0.0041 & 0.0529 & 0.0201 & 0.0201 \\ 0.0088 & -0.0090 & 0.0201 & 0.0625 & -0.0131 \\ -0.0098 & 0.0095 & 0.0201 & -0.0131 & 0.1225 \end{pmatrix}. \quad (9.17)$$

(i) The minimum variance portfolio can be found by solving a constrained optimization problem using Lagrange multipliers. If $w = (w_i)_{i=1:5}$ is the vector of the weights of the assets in the minimum variance portfolio, recall that w can be obtained by solving the following linear system:

$$\begin{pmatrix} 2M & \mathbf{1} & \mu \\ \mathbf{1}^t & 0 & 0 \\ \mu^t & 0 & 0 \end{pmatrix} \begin{pmatrix} w \\ \lambda_1 \\ \lambda_2 \end{pmatrix} = \begin{pmatrix} 0 \\ 1 \\ \mu_P \end{pmatrix}, \quad (9.18)$$

where $\mu_P = 0.15$ is the required expected rate of return of the minimum variance portfolio.

For M given by (9.17) and μ given by (9.15), the solution of the linear system (9.18) is

$$\begin{pmatrix} w_{0,1} \\ w_{0,2} \\ w_{0,3} \\ w_{0,4} \\ w_{0,5} \\ \lambda_{0,1} \\ \lambda_{0,2} \end{pmatrix} = \begin{pmatrix} 0.067482 \\ 0.217330 \\ -0.002571 \\ 0.406983 \\ 0.310776 \\ 0.044892 \\ -0.565252 \end{pmatrix}.$$

The asset allocation for a minimum variance portfolio with 15% expected rate of return is as follows:

- invest 6.75% of the value of the portfolio in asset 1;
- invest 21.73% of the value of the portfolio in asset 2;
- short an amount of asset 3 equal to 0.26% of the value of the portfolio;
- invest 40.70% of the value of the portfolio in asset 4;
- invest 31.08% of the value of the portfolio in asset 5.

For example, if the value of the portfolio is $10,000,000, then $341,000 of asset 1 is shorted (borrowed and sold for cash) $2,603,000 is invested in asset 2, $839,000 is invested in asset 3, $4,439,000 is invested in asset 4, and $2,461,000 is invested in asset 5.
(Due to roundoff errors, $-341,000 + 2,603,000 + 839,000 + 4,439,000 + 2,461,000 = 10,001,000$).

The minimum variance portfolio has a standard deviation of the expected rate of return equal to 15.35%. (Recall that the standard deviation of the portfolio is $\sigma_P = \sqrt{w_0^t M w_0}$.)

(ii) Recall that the vector μ of the expected rates of return of the assets and the covariance matrix M of the rates of return of the assets are given by (9.15) and (9.17), respectively.

We use the Lagrange multipliers method to find the maximum return portfolio with standard deviation of the rate of return $\sigma_P = 0.25$. We first check that

$$\mathbf{1}^t M^{-1} \mathbf{1} = 139.27 \neq 16 = \frac{1}{\sigma_P^2}.$$

The weights $w = (w_i)_{i=1:5}$ of the maximum return portfolio can be found by solving

$$G(w, \lambda_1, \lambda_2) = 0,$$

where the function $G(w, \lambda_1, \lambda_2) : \mathbb{R}^7 \to \mathbb{R}^7$ is given by[2]

$$G(w, \lambda_1, \lambda_2) = \begin{pmatrix} \mu + \lambda_1 \mathbf{1} + 2\lambda_2 M w \\ \mathbf{1}^t w - 1 \\ w^t M w - \sigma_P^2 \end{pmatrix}.$$

The solution to $G(w, \lambda_1, \lambda_2) = 0$ is obtained numerically, using a seven dimensional Newton's method. The gradient of $G(w, \lambda_1, \lambda_2)$ is the following 7×7 matrix:

$$\nabla_{(w,\lambda)} G(w, \lambda_1, \lambda_2) = \begin{pmatrix} 2\lambda_2 M & \mathbf{1} & 2Mw \\ \mathbf{1}^t & 0 & 0 \\ 2(Mw)^t & 0 & 0 \end{pmatrix}.$$

The corresponding Newton's method recursion is

$$x_{k+1} = x_k - (\nabla G(x_k))^{-1} G(x_k), \tag{9.19}$$

where x_k is a seven dimensional column vector whose first five entries correspond to the portfolio weights of the five assets and whose last two entries correspond to the Lagrange multipliers. The solution to the Newton's method iteration (9.19) with initial guess

$$x_0 = \begin{pmatrix} 0.5 \\ 0.5 \\ 0.5 \\ 0.5 \\ 0.5 \\ 1 \\ 1 \end{pmatrix} \quad \text{is} \quad x^* = \begin{pmatrix} -0.4794 \\ 0.1864 \\ 0.1767 \\ 0.7173 \\ 0.3990 \\ -0.0899 \\ -0.8057 \end{pmatrix}.$$

[2]Recall that $G(w, \lambda_1, \lambda_2)$ is the transpose of the gradient of the Lagrangian function associated to this constrained optimization problem.

In other words, the Lagrangian corresponding to this problem has one critical point given by

$$w_0 = \begin{pmatrix} -0.4794 \\ 0.1864 \\ 0.1767 \\ 0.7173 \\ 0.3990 \end{pmatrix}; \quad \lambda_{0,1} = -0.0899; \lambda_{0,2} = -0.8057.$$

Since $\lambda_{0.2} < 0$, the point w_0 is a constrained maximum point.

We conclude that the portfolio with 25% standard deviation of the rate of return and maximal expected return is obtained as follows:

- short an amount of asset 1 equal to 47.94% of the value of the portfolio;
- invest 18.64% of the value of the portfolio in asset 2;
- invest 17.67% of the value of the portfolio in asset 3;
- invest 71.73% of the value of the portfolio in asset 4;
- invest 39.90% of the value of the portfolio in asset 5.

For example, if the value of the portfolio is $10,000,000, then $4,794,000 of asset 1 is shorted (borrowed and sold for cash), $1,864,000 is invested in asset 2, $1,767,000 is invested in asset 3, $7,173,000 is invested in asset 4, and $3,990,000 is invested in asset 5.

The expected rate of return of this portfolio is 19.06%. (Recall that the expected rate of return of the portfolio is $\mu_P = \mu^t w_0$.) □

Chapter 10

Mathematical Appendix

10.1 Exercises

1. Let $f : \mathbb{R} \to \mathbb{R}$ be an odd function.

   (i) Show that $xf(x)$ is an even function and $x^2 f(x)$ is an odd function.

   (ii) Show that the function $g_1 : \mathbb{R} \to \mathbb{R}$ given by $g_1(x) = f(x^2)$ is an even function and that the function $g_2 : \mathbb{R} \to \mathbb{R}$ given by $g_2(x) = f(x^3)$ is an odd function.

   (iii) More generally, let $h : \mathbb{R} \to \mathbb{R}$ be defined as

   $$h(x) = x^i f(x^j),$$

   where i and j are positive integers. Show that $h(x)$ is an odd function if $i + j$ is an odd integer, and $h(x)$ is an even function if $i + j$ is an even integer.

2. Let

   $$S(n,2) = \sum_{k=1}^{n} k^2; \quad S(n,3) = \sum_{k=1}^{n} k^3; \quad T(n,2,x) = \sum_{k=1}^{n} k^2 \, x^k.$$

   Recall that

   $$T(n,2,x) = \frac{x + x^2 - (n+1)^2 x^{n+1} + (2n^2 + 2n - 1)x^{n+2} - n^2 x^{n+3}}{(1-x)^3}.$$

   (i) Note that $S(n,2) = T(n,2,1)$. Evaluate $T(n,2,1)$ using l'Hôpital's rule and conclude that

   $$S(n,2) = \frac{n(n+1)(2n+1)}{6}.$$

(ii) Compute $T(n,3,x) = \sum_{k=1}^n k^3 \, x^k$ using the fact that

$$T(n,3,x) \; = \; x \, \frac{d}{dx}(T(n,2,x)).$$

(iii) Note that $S(n,3) = T(n,3,1)$. Evaluate $T(n,3,1)$ using l'Hôpital's rule and conclude that

$$S(n,3) \; = \; \left(\frac{n(n+1)}{2} \right)^2.$$

3. Compute $S(n,4) = \sum_{k=1}^n k^4$ using the recursion formula

$$S(n,i) \; = \; \frac{1}{i+1} \left((n+1)^{i+1} - 1 \; - \; \sum_{j=0}^{i-1} \binom{i+1}{j} \; S(n,j) \right)$$

for $i = 4$, given that

$$S(n,0) \; = \; n; \quad S(n,1) \; = \; \frac{n(n+1)}{2}; \quad S(n,2) \; = \; \frac{n(n+1)(2n+1)}{6};$$

$$S(n,3) \; = \; \left(\frac{n(n+1)}{2} \right)^2.$$

4. It is easy to see that the sequence $(x_n)_{n \geq 1}$ given by $x_n = \sum_{k=1}^n k^2$ satisfies the recursion

$$x_{n+1} \; = \; x_n + (n+1)^2, \quad \forall \, n \geq 1, \tag{10.1}$$

with $x_1 = 1$.

(i) By substituting $n+1$ for n in (10.1), obtain that

$$x_{n+2} \; = \; x_{n+1} + (n+2)^2. \tag{10.2}$$

Subtract (10.1) from (10.2) to find that

$$x_{n+2} \; = \; 2x_{n+1} - x_n + 2n + 3, \quad \forall \, n \geq 1, \tag{10.3}$$

with $x_1 = 1$ and $x_2 = 5$.

(ii) Similarly, substitute $n+1$ for n in (10.3) and obtain that

$$x_{n+3} \; = \; 2x_{n+2} - x_{n+1} + 2(n+1) + 3. \tag{10.4}$$

Subtract (10.3) from (10.4) to find that

$$x_{n+3} = 3x_{n+2} - 3x_{n+1} + x_n + 2, \quad \forall\, n \geq 1,$$

with $x_1 = 1$, $x_2 = 5$, and $x_3 = 14$.

(iii) Use a similar method to prove that the sequence $(x_n)_{n\geq 1}$ satisfies the linear recursion

$$x_{n+4} - 4x_{n+3} + 6x_{n+2} - 4x_{n+1} + x_n = 0, \quad \forall\, n \geq 1. \tag{10.5}$$

The characteristic polynomial associated to the recursion (10.5) is

$$P(z) = z^4 - 4z^3 + 6z^2 - 4z + 1 = (z-1)^4.$$

Use the fact that $x_1 = 1$, $x_2 = 5$, $x_3 = 14$, and $x_4 = 30$ to show that

$$x_n = \frac{n(n+1)(2n+1)}{6}, \quad \forall\, n \geq 1,$$

and conclude that

$$S(n,2) = \sum_{k=1}^{n} k^2 = \frac{n(n+1)(2n+1)}{6}, \quad \forall\, n \geq 1.$$

5. Find the general form of the sequence $(x_n)_{n\geq 0}$ satisfying the linear recursion

$$x_{n+3} = 2x_{n+1} + x_n, \quad \forall\, n \geq 0,$$

with $x_0 = 1$, $x_1 = -1$, and $x_2 = 1$.

6. The sequence $(x_n)_{n\geq 0}$ satisfies the recursion

$$x_{n+1} = 3x_n + 2, \quad \forall\, n \geq 0,$$

with $x_0 = 1$.

(i) Show that the sequence $(x_n)_{n\geq 0}$ satisfies the linear recursion

$$x_{n+2} = 4x_{n+1} - 3x_n, \quad \forall\, n \geq 0,$$

with $x_0 = 1$ and $x_1 = 5$.

(ii) Find the general formula for x_n, $n \geq 0$.

7. The sequence $(x_n)_{n\geq 0}$ satisfies the recursion

$$x_{n+1} = 3x_n + n + 2, \quad \forall\, n \geq 0,$$

with $x_0 = 1$.

(i) Show that the sequence $(x_n)_{n\geq 0}$ satisfies the linear recursion

$$x_{n+3} = 5x_{n+2} - 7x_{n+1} + 3x_n, \quad \forall\, n \geq 0,$$

with $x_0 = 1$, $x_1 = 5$, and $x_2 = 18$.

(ii) Find the general formula for x_n, $n \geq 0$.

8. Let $P(z) = \sum_{i=0}^{k} a_i z^i$ be the characteristic polynomial corresponding to the linear recursion

$$\sum_{i=0}^{k} a_i x_{n+i} = 0, \quad \forall\, n \geq 0. \tag{10.6}$$

Assume that λ is a root of multiplicity 2 of $P(z)$. Show that the sequence $(y_n)_{n\geq 0}$ given by

$$y_n = Cn\lambda^n, \quad n \geq 0,$$

where C is an arbitrary constant, satisfies the recursion (10.6).

Hint: Show that

$$\sum_{i=0}^{k} a_i y_{n+i} = Cn\lambda^n P(\lambda) \;+\; C\lambda^{n+1} P'(\lambda), \quad \forall\, n \geq 0,$$

and recall that λ is a root of multiplicity 2 of the polynomial $P(z)$ if and only if $P(\lambda) = 0$ and $P'(\lambda) = 0$.

9. Let $n > 0$. Show that

$$\begin{aligned} O(x^n) + O(x^n) &= O(x^n), \quad \text{as} \quad x \to 0; \qquad (10.7)\\ o(x^n) + o(x^n) &= o(x^n), \quad \text{as} \quad x \to 0. \end{aligned}$$

For example, to prove (10.7), let $f(x) = O(x^n)$ and $g(x) = O(x^n)$ as $x \to 0$, and show that $f(x) + g(x) = O(x^n)$ as $x \to 0$, i.e., that

$$\limsup_{x\to 0} \left| \frac{f(x) + g(x)}{x^n} \right| < \infty.$$

10. Prove that

$$\sum_{k=1}^{n} k^2 = O(n^3), \quad \text{as} \quad n \to \infty;$$

$$\sum_{k=1}^{n} k^2 = \frac{n^3}{3} + O(n^2), \quad \text{as} \quad n \to \infty,$$

i.e., show that

$$\limsup_{n\to\infty} \frac{\sum_{k=1}^{n} k^2}{n^3} < \infty$$

and that

$$\limsup_{n\to\infty} \frac{\sum_{k=1}^{n} k^2 - \frac{n^3}{3}}{n^2} < \infty.$$

Similarly, prove that

$$\sum_{k=1}^{n} k^3 = O(n^4), \quad \text{as} \quad n \to \infty;$$

$$\sum_{k=1}^{n} k^3 = \frac{n^4}{4} + O(n^3), \quad \text{as} \quad n \to \infty.$$

11. Use the binomial formula

$$(a+b)^n = \sum_{j=0}^{n} \binom{n}{j} a^j b^{n-j},$$

for all $a, b \in \mathbb{R}$, and for any positive integer $n \geq 1$, to establish the following equalities:

$$\sum_{j=0}^{n} \binom{n}{j} = 2^n;$$

$$\sum_{k=0}^{\lfloor \frac{n}{2} \rfloor} \binom{n}{2k} = 2^{n-1};$$

$$\sum_{j=0}^{n} j \binom{n}{j} = n2^{n-1}.$$

12. Show that the series

$$\sum_{k=1}^{\infty} \frac{1}{k^2}$$

is convergent, while the series

$$\sum_{k=1}^{\infty} \frac{1}{k} \quad \text{and} \quad \sum_{k=2}^{\infty} \frac{1}{k \ln(k)}$$

are divergent, i.e., equal to ∞.

Note: It is known that

$$\sum_{k=1}^{\infty} \frac{1}{k^2} = \frac{\pi^2}{6}$$

and

$$\lim_{n \to \infty} \left(\sum_{k=1}^{\infty} \frac{1}{k} - \ln(n) \right) = \gamma,$$

where $\gamma \approx 0.57721$ is called Euler's constant.

Hint: Note that

$$\sum_{k=1}^{n-1} \frac{1}{k+1} < \ln(n) = \int_1^n \frac{1}{x}\, dx = \sum_{k=1}^{n-1} \int_k^{k+1} \frac{1}{x}\, dx < \sum_{k=1}^{n-1} \frac{1}{k},$$

since $\frac{1}{k+1} < \frac{1}{x} < \frac{1}{k}$, for any $k < x < k+1$. Conclude that

$$\ln(n) + \frac{1}{n} < \sum_{k=1}^{n} \frac{1}{k} < \ln(n) + 1, \quad \forall\, n \geq 1;$$

For the series $\sum_{k=2}^{\infty} \frac{1}{k \ln(k)}$, use a similar method to find upper and lower bounds for the integral of $\frac{1}{x \ln(x)}$ over the interval $[2, n]$.

13. Show that the sequence

$$x_n = \left(\sum_{k=1}^{n} \frac{1}{k} \right) - \ln(n)$$

is convergent to a limit between 0 and 1.

Note: The limit of this sequence is $\gamma \approx 0.57721$, the Euler's constant.

14. Find the radius of convergence R of the power series

$$\sum_{k=2}^{\infty} \frac{x^k}{k \ \ln(k)},$$

and investigate what happens at the points x where $|x| = R$.

15. Let $a > 0$ be a positive number. Compute

$$\sqrt{a + \sqrt{a + \sqrt{a + \ldots}}}$$

16. Let $a > 0$ be a positive number. Compute

$$a + \frac{1}{a + \frac{1}{a + \ldots}}.$$

17. (i) Find $x > 0$ such that

$$x^{x^{x^{\cdot^{\cdot^{\cdot}}}}} = 2.$$

(ii) Find the largest possible value of $x > 0$ with such that there exists a number $b > 0$ with

$$x^{x^{x^{\cdot^{\cdot^{\cdot}}}}} = b.$$

Also, what is the largest possible value of b?

10.2 Solutions to Chapter 10 Exercises

Problem 1: Let $f : \mathbb{R} \to \mathbb{R}$ be an odd function.
(i) Show that $xf(x)$ is an even function and $x^2 f(x)$ is an odd function.
(ii) Show that the function $g_1 : \mathbb{R} \to \mathbb{R}$ given by $g_1(x) = f(x^2)$ is an even function and that the function $g_2 : \mathbb{R} \to \mathbb{R}$ given by $g_2(x) = f(x^3)$ is an odd function.
(iii) More generally, let $h : \mathbb{R} \to \mathbb{R}$ be defined as

$$h(x) = x^i f(x^j),$$

where i and j are positive integers. Show that $h(x)$ is an odd function if $i+j$ is an odd integer, and $h(x)$ is an even function if $i+j$ is an even integer.

Solution: Since $f(x)$ is an odd function, it follows that

$$f(-x) = -f(x), \quad \forall\, x \in \mathbb{R}. \tag{10.8}$$

(i) Let $f_1(x) = xf(x)$ and $f_2(x) = x^2 f(x)$. Using (10.8), we find that

$$f_1(-x) = -xf(-x) = xf(x) = f_1(x), \quad \forall\, x \in \mathbb{R}; \tag{10.9}$$
$$f_2(-x) = (-x)^2 f(-x) = -x^2 f(x) = -f_2(x), \quad \forall\, x \in \mathbb{R}. \tag{10.10}$$

We conclude from (10.9) that $f_1(x)$ is an even function, and, from (10.10), $f_2(x)$ is an odd function.
(ii) From (10.8), it follows that

$$g_1(-x) = f\left((-x)^2\right) = f(x^2) = g_1(x), \quad \forall\, x \in \mathbb{R}; \tag{10.11}$$
$$g_2(-x) = f\left((-x)^3\right) = f(-x^3) = -f(x^3) = -g_2(x), \quad \forall\, x \in \mathbb{R} \tag{10.12}$$

We conclude from (10.11) that $g_1(x)$ is an even function, and, from (10.12), that $g_2(x)$ is an odd function.
(iii) If j is a positive integer, it is easy to see from (10.8) that

$$f\left((-x)^j\right) = (-1)^j f(x^j), \quad \forall\, x \in \mathbb{R}. \tag{10.13}$$

Then, using (10.13), we find that

$$\begin{aligned} h(-x) &= (-x)^i f\left((-x)^j\right) = (-1)^i x^i \cdot (-1)^j f(x^j) = (-1)^{i+j} x^i f(x^j) \\ &= (-1)^{i+j} h(x), \quad \forall\, x \in \mathbb{R}. \end{aligned}$$

Therefore, if $i+j$ is an even integer, the function $h(x)$ is an even function, and, if $i+j$ is an odd integer, the function $h(x)$ is an odd function. □

Problem 2: Let

$$S(n,2) = \sum_{k=1}^{n} k^2; \quad S(n,3) = \sum_{k=1}^{n} k^3; \quad T(n,2,x) = \sum_{k=1}^{n} k^2 \, x^k.$$

Recall that

$$T(n,2,x) = \frac{x + x^2 - (n+1)^2 x^{n+1} + (2n^2+2n-1)x^{n+2} - n^2 x^{n+3}}{(1-x)^3}.$$

(i) Note that $S(n,2) = T(n,2,1)$. Use l'Hôpital's rule to evaluate $T(n,2,1)$, and conclude that

$$S(n,2) = \frac{n(n+1)(2n+1)}{6}.$$

(ii) Compute $T(n,3,x) = \sum_{k=1}^{n} k^3 \, x^k$ using the recursion formula

$$T(n,3,x) = x \frac{d}{dx}(T(n,2,x)).$$

(iii) Note that $S(n,3) = T(n,3,1)$. Use l'Hôpital's rule to evaluate $T(n,3,1)$, and conclude that

$$S(n,3) = \left(\frac{n(n+1)}{2}\right)^2.$$

Solution: (i) It is easy to see that $T(n,2,1) = \sum_{k=1}^{n} k^2 = S(n,2)$. By using l'Hôpital's rule we find that $T(n,2,x)$ is equal to

$$\begin{aligned}
& \lim_{x\to 1} \frac{x + x^2 - (n+1)^2 x^{n+1} + (2n^2+2n-1)x^{n+2} - n^2 x^{n+3}}{(1-x)^3} \\
= & \lim_{x\to 1} \frac{1 + 2x - (n+1)^3 x^n + (2n^2+2n-1)(n+2)x^{n+1} - n^2(n+3)x^{n+2}}{-3(1-x)^2} \\
= & \lim_{x\to 1} \frac{\left(\begin{array}{l} 2 - (n+1)^3 n x^{n-1} + (2n^2+2n-1)(n+2)(n+1)x^n \\ -n^2(n+3)(n+2)x^{n+1} \end{array}\right)}{6(1-x)} \\
= & \lim_{x\to 1} \frac{\left(\begin{array}{l} -(n+1)^3 n(n-1)x^{n-2} + (2n^2+2n-1)(n+2)(n+1)n x^{n-1} \\ -n^2(n+3)(n+2)(n+1)x^n \end{array}\right)}{-6} \\
= & -\frac{\left(\begin{array}{l} -(n+1)^3 n(n-1) + (2n^2+2n-1)(n+2)(n+1)n \\ -n^2(n+3)(n+2)(n+1) \end{array}\right)}{6} \\
= & -\frac{n(n+1)\left(\begin{array}{l} -(n+1)^2(n-1) + (2n^2+2n-1)(n+2) \\ -n(n+3)(n+2) \end{array}\right)}{6}
\end{aligned}$$

$$= \frac{n(n+1)(2n+1)}{6}.$$

Therefore,

$$S(n,2) = \frac{n(n+1)(2n+1)}{6}.$$

(ii) Finding the value of $T(n,3,x)$ requires using Quotient Rule to differentiate $T(n,2,x)$.

(iii) The solution follows similarly to that from part (i), albeit with more complicated computations. □

Problem 3: Compute $S(n,4) = \sum_{k=1}^{n} k^4$ using the recursion formula

$$S(n,i) = \frac{1}{i+1}\left((n+1)^{i+1} - 1 - \sum_{j=0}^{i-1}\binom{i+1}{j} S(n,j)\right) \tag{10.14}$$

for $i = 4$, given that

$$S(n,0) = n; \quad S(n,1) = \frac{n(n+1)}{2}; \quad S(n,2) = \frac{n(n+1)(2n+1)}{6};$$

$$S(n,3) = \left(\frac{n(n+1)}{2}\right)^2.$$

Solution: For $i = 4$, the recursion formula (10.14) becomes

$$\begin{aligned} S(n,4) &= \frac{1}{5}\left((n+1)^5 - 1 - \sum_{j=0}^{3}\binom{5}{j} S(n,j)\right) \\ &= \frac{1}{5}\left((n+1)^5 - 1 - S(n,0) - 5S(n,1) - 10S(n,2) - 10S(n,3)\right) \\ &= \frac{n(n+1)(6n^3 + 9n^2 + n - 1)}{30}. \quad \square \end{aligned}$$

Problem 4: It is easy to see that the sequence $(x_n)_{n\geq 1}$ given by $x_n = \sum_{k=1}^{n} k^2$ satisfies the recursion

$$x_{n+1} = x_n + (n+1)^2, \quad \forall\, n \geq 1, \tag{10.15}$$

with $x_1 = 1$.

(i) By substituting $n+1$ for n in (10.15), obtain that

$$x_{n+2} \;=\; x_{n+1} + (n+2)^2. \tag{10.16}$$

Subtract (10.15) from (10.16) to find that

$$x_{n+2} \;=\; 2x_{n+1} - x_n + 2n + 3, \quad \forall\, n \geq 1, \tag{10.17}$$

with $x_1 = 1$ and $x_2 = 5$.
(ii) Similarly, show that

$$x_{n+3} \;=\; 3x_{n+2} - 3x_{n+1} + x_n + 2, \quad \forall\, n \geq 1, \tag{10.18}$$

with $x_1 = 1$, $x_2 = 5$, and $x_3 = 14$.
(iii) Prove that the sequence $(x_n)_{n\geq 1}$ satisfies the linear recursion

$$x_{n+4} - 4x_{n+3} + 6x_{n+2} - 4x_{n+1} + x_n \;=\; 0, \quad \forall\, n \geq 1.$$

Solve this recursion and show that

$$x_n \;=\; \frac{n(n+1)(2n+1)}{6}, \quad \forall\, n \geq 1.$$

Conclude that

$$S(n,2) \;=\; \sum_{k=1}^{n} k^2 \;=\; \frac{n(n+1)(2n+1)}{6}, \quad \forall\, n \geq 1.$$

Solution: From (10.16), we obtain that the first terms of the sequence $(x_n)_{n\geq 1}$ are $x_1 = 1$, $x_2 = 5$, $x_3 = 14$, $x_4 = 30$.
(i) The recursion (10.17) can be obtained by subtracting (10.15) from (10.16).
(ii) We substitute $n+1$ for n in (10.17) and obtain that

$$x_{n+3} \;=\; 2x_{n+2} - x_{n+1} + 2(n+1) + 3. \tag{10.19}$$

By subtracting (10.17) from (10.19), we find that

$$x_{n+3} \;=\; 3x_{n+2} - 3x_{n+1} + x_n + 2, \quad \forall\, n \geq 1.$$

(iii) We substitute $n+1$ for n in (10.18) and obtain that

$$x_{n+4} \;=\; 3x_{n+1} - 3x_{n+2} + x_{n+1} + 2, \quad \forall\, n \geq 1. \tag{10.20}$$

By subtracting (10.18) from (10.20), we find that

$$x_{n+4} - 4x_{n+3} + 6x_{n+2} - 4x_{n+1} + x_n \;=\; 0, \quad \forall\, n \geq 1. \tag{10.21}$$

The characteristic polynomial associated to the recursion (10.21) is

$$P(z) \;=\; z^4 - 4z^3 + 6z^2 - 4z + 1 \;=\; (z-1)^4.$$

The polynomial $P(z)$ has root $\lambda = 1$ with multiplicity 4. We conclude that the there exist constants C_i, $i = 1:4$, such that

$$x_n \;=\; C_1 + C_2 n + C_3 n^2 + C_4 n^3, \quad \forall\, n \geq 1.$$

Since $x_1 = 1$, $x_2 = 5$, $x_3 = 14$, $x_4 = 30$, it follows that C_1, C_2, C_3 and C_4 satisfy the following linear system

$$\begin{pmatrix} 1 & 1 & 1 & 1 \\ 1 & 2 & 4 & 8 \\ 1 & 3 & 9 & 27 \\ 1 & 4 & 16 & 64 \end{pmatrix} \begin{pmatrix} C_1 \\ C_2 \\ C_3 \\ C_4 \end{pmatrix} \;=\; \begin{pmatrix} 1 \\ 5 \\ 14 \\ 30 \end{pmatrix}.$$

We obtain that $C_1 = 0$, $C_2 = \frac{1}{6}$, $C_3 = \frac{1}{2}$ and $C_4 = \frac{1}{3}$ and therefore

$$x_n \;=\; \frac{n}{6} + \frac{n^2}{2} + \frac{n^3}{3} \;=\; \frac{n(n+1)(2n+1)}{6}, \quad \forall\, n \geq 1. \quad \square$$

Problem 5: Find the general form of the sequence $(x_n)_{n \geq 0}$ satisfying the linear recursion

$$x_{n+3} \;=\; 2x_{n+1} \;+\; x_n, \quad \forall\, n \geq 0,$$

with $x_0 = 1$, $x_1 = -1$, and $x_2 = 1$.

Solution 1: By direct computation, we obtain $x_3 = -1$, $x_4 = 1$, $x_5 = -1$, $x_6 = 1$. It is natural to conjecture that $x_n = (-1)^n$ for any positive integer n. This can be easily checked by induction.

Solution 2: Alternatively, we note that the sequence $(x_n)_{n \geq 0}$ satisfies the linear recursion $x_{n+3} - 2x_{n+1} - x_n = 0$, with characteristic polynomial

$$P(z) \;=\; z^3 - 2z - 1 \;=\; (z+1)(z^2 - z - 1).$$

The roots of $P(z)$ are -1, $\frac{1+\sqrt{5}}{2}$, and $\frac{1-\sqrt{5}}{2}$. Therefore, there exist constants C_1, C_2, and C_3 such that

$$x_n \;=\; C_1(-1)^n \;+\; C_2\left(\frac{1+\sqrt{5}}{2}\right)^n \;+\; C_3\left(\frac{1-\sqrt{5}}{2}\right)^n, \quad \forall\, n \geq 0.$$

By solving the 3×3 linear system for C_1, C_2, and C_3 obtained by requiring that $x_0 = 1$, $x_1 = -1$, and $x_2 = 1$ we find that $C_1 = 1$, $C_2 = 0$, and $C_3 = 0$. We conclude that

$$x_n \;=\; (-1)^n, \quad \forall\, n \geq 0. \quad \square$$

Problem 6: The sequence $(x_n)_{n \geq 0}$ satisfies the recursion

$$x_{n+1} = 3x_n + 2, \quad \forall\, n \geq 0, \tag{10.22}$$

with $x_0 = 1$.
(i) Show that the sequence $(x_n)_{n \geq 0}$ satisfies the linear recursion

$$x_{n+2} = 4x_{n+1} - 3x_n, \quad \forall\, n \geq 0,$$

with $x_0 = 1$ and $x_1 = 5$.
(ii) Find the general formula for x_n, $n \geq 0$.

Solution: (i) By letting $n = 0$ in (10.22), we find that $x_1 = 5$.
By substituting $n + 1$ for n in (10.22), it follows that

$$x_{n+2} = 3x_{n+1} + 2, \quad \forall\, n \geq 0. \tag{10.23}$$

We subtract (10.22) from (10.23) and obtain that

$$x_{n+2} - 4x_{n+1} + 3x_n = 0, \quad \forall\, n \geq 0. \tag{10.24}$$

(ii) The characteristic polynomial of the linear recursion (10.24) is

$$P(z) = z^2 - 4z + 3 = (z-1)(z-3),$$

which has roots 1 and 3. Thus,

$$x_n = C_1 + C_2 \cdot 3^n, \quad \forall\, n \geq 0.$$

Since $x_0 = 1$ and $x_1 = 5$, we obtain that $C_1 = -1$ and $C_2 = 2$ and therefore

$$x_n = 2 \cdot 3^n - 1, \quad \forall\, n \geq 0. \quad \square$$

Problem 7: The sequence $(x_n)_{n \geq 0}$ satisfies the recursion

$$x_{n+1} = 3x_n + n + 2, \quad \forall\, n \geq 0, \tag{10.25}$$

with $x_0 = 1$.
(i) Show that the sequence $(x_n)_{n \geq 0}$ satisfies the linear recursion

$$x_{n+3} = 5x_{n+2} - 7x_{n+1} + 3x_n, \quad \forall\, n \geq 0,$$

with $x_0 = 1$, $x_1 = 5$, and $x_2 = 18$.
(ii) Find the general formula for x_n, $n \geq 0$.

Solution: (i) The first three terms of the sequence can be computed from (10.25) and are $x_0 = 1$, $x_1 = 5$, and $x_2 = 18$.

By substituting $n+1$ for n in (10.25) we obtain that

$$x_{n+2} = 3x_{n+1} + n + 3, \quad \forall\, n \geq 0. \qquad (10.26)$$

Subtract (10.25) from (10.26) to find that

$$x_{n+2} = 4x_{n+1} - 3x_n + 1, \quad \forall\, n \geq 0. \qquad (10.27)$$

Substitute $n+1$ for n in (10.27) to obtain that

$$x_{n+3} = 4x_{n+2} - 3x_{n+1} + 1, \quad \forall\, n \geq 0. \qquad (10.28)$$

Subtract (10.27) from (10.28) to find that

$$x_{n+3} = 5x_{n+2} - 7x_{n+1} + 3x_n, \quad \forall\, n \geq 0. \qquad (10.29)$$

(ii) The characteristic polynomial of the linear recursion (10.29) is

$$P(x) = z^3 - 5z^2 + 7z - 3 = (z-1)^2(z-3).$$

Therefore, there exist constants C_1, C_2, C_3 such that

$$x_n = C_1 3^n + C_2 n + C_3, \quad \forall\, n \geq 0.$$

Since $x_0 = 1$, $x_1 = 5$, and $x_2 = 18$, we find that $C_1 = \frac{9}{4}$, $C_2 = -\frac{5}{4}$, $C_3 = -\frac{1}{2}$. We conclude that

$$x_n = \frac{3^{n+2} - 2n - 5}{4}, \quad \forall\, n \geq 0. \quad \Box$$

Problem 8: Let $P(z) = \sum_{i=0}^{k} a_i z^i$ be the characteristic polynomial corresponding to the linear recursion

$$\sum_{i=0}^{k} a_i x_{n+i} = 0, \quad \forall\, n \geq 0. \qquad (10.30)$$

Assume that λ is a root of multiplicity 2 of $P(z)$. Show that the sequence $(y_n)_{n\geq 0}$ given by

$$y_n = Cn\lambda^n, \quad n \geq 0,$$

where C is an arbitrary constant, satisfies the recursion (10.30).

Solution: Note that λ is a root of multiplicity 2 of $P(z)$ if and only if $P(\lambda) = 0$ and $P'(\lambda) = 0$.

Then, for any $n \geq 0$,

$$\begin{aligned}
\sum_{i=0}^{k} a_i y_{n+i} &= \sum_{i=0}^{k} a_i C(n+i)\lambda^{n+i} \\
&= Cn\sum_{i=0}^{k} a_i\lambda^{n+i} + C\sum_{i=0}^{k} ia_i\lambda^{n+i} \\
&= Cn\lambda^n\sum_{i=0}^{k} a_i\lambda^i + C\lambda^{n+1}\sum_{i=1}^{k} ia_i\lambda^{i-1} \\
&= Cn\lambda^n P(\lambda) + C\lambda^{n+1}P'(\lambda) \\
&= 0,
\end{aligned}$$

since

$$P'(z) = \sum_{i=1}^{k} ia_i z^{i-1}.$$

In other words, the sequence $(y_n)_{n\geq 0}$ satisfies the recursion (10.30). □

Problem 9: Let $n > 0$. Show that

$$O(x^n) + O(x^n) = O(x^n), \quad \text{as} \quad x \to 0; \tag{10.31}$$

$$o(x^n) + o(x^n) = o(x^n), \quad \text{as} \quad x \to 0. \tag{10.32}$$

Solution: Let $f_1(x) = O(x^n)$ and $f_2(x) = O(x^n)$ as $x \to 0$. Then,

$$\limsup_{x\to 0}\left|\frac{f_1(x)}{x^n}\right| < \infty \quad \text{and} \quad \limsup_{x\to 0}\left|\frac{f_2(x)}{x^n}\right| < \infty.$$

It is easy to see that

$$\limsup_{x\to 0}\left|\frac{f_1(x)+f_2(x)}{x^n}\right| \leq \limsup_{x\to 0}\left|\frac{f_1(x)}{x^n}\right| + \limsup_{x\to 0}\left|\frac{f_2(x)}{x^n}\right| < \infty,$$

and therefore, by definition,

$$f_1(x) + f_2(x) = O(x^n), \quad \text{as} \quad x \to 0.$$

Let $g_1(x) = o(x^n)$ and $g_2(x) = o(x^n)$ as $x \to 0$. Then,

$$\lim_{x\to 0}\left|\frac{g_1(x)}{x^n}\right| = 0 \quad \text{and} \quad \lim_{x\to 0}\left|\frac{g_2(x)}{x^n}\right| = 0.$$

We note that

$$\lim_{x\to 0}\left|\frac{g_1(x)+g_2(x)}{x^n}\right| \le \lim_{x\to 0}\left|\frac{g_1(x)}{x^n}\right| + \lim_{x\to 0}\left|\frac{g_2(x)}{x^n}\right| = 0,$$

and therefore, by definition,

$$g_1(x)+g_2(x) = o(x^n), \quad \text{as} \quad x \to 0. \quad \square$$

Problem 10: Prove that

$$\sum_{k=1}^{n} k^2 = O(n^3), \quad \text{as} \quad n \to \infty;$$

$$\sum_{k=1}^{n} k^2 = \frac{n^3}{3} + O(n^2), \quad \text{as} \quad n \to \infty,$$

i.e., show that

$$\limsup_{n\to\infty} \frac{\sum_{k=1}^{n} k^2}{n^3} < \infty$$

and that

$$\limsup_{n\to\infty} \frac{\sum_{k=1}^{n} k^2 - \frac{n^3}{3}}{n^2} < \infty.$$

Similarly, prove that

$$\sum_{k=1}^{n} k^3 = O(n^4), \quad \text{as} \quad n \to \infty;$$

$$\sum_{k=1}^{n} k^3 = \frac{n^4}{4} + O(n^3), \quad \text{as} \quad n \to \infty.$$

Solution: Recall that

$$\sum_{k=1}^{n} k^2 = \frac{n(n+1)(2n+1)}{6} \quad \text{and} \quad \sum_{k=1}^{n} k^3 = \frac{n^2(n+1)^2}{4}.$$

Then,

$$\lim_{n\to\infty} \frac{\sum_{k=1}^{n} k^2}{n^3} = \lim_{n\to\infty} \frac{n(n+1)(2n+1)}{6n^3} = \frac{1}{3} < \infty;$$

$$\lim_{n\to\infty} \frac{\sum_{k=1}^{n} k^2 - \frac{n^3}{3}}{n^2} = \lim_{n\to\infty} \frac{3n^2+n}{6n^2} = \frac{1}{2} < \infty;$$

$$\lim_{n\to\infty} \frac{\sum_{k=1}^{n} k^3}{n^4} = \lim_{n\to\infty} \frac{n^2(n+1)^2}{4n^4} = \frac{1}{4} < \infty;$$

$$\lim_{n\to\infty} \frac{\sum_{k=1}^{n} k^3 - \frac{n^4}{4}}{n^3} = \lim_{n\to\infty} \frac{2n+1}{4n} = \frac{1}{2} < \infty.$$

We conclude that

$$\begin{aligned}
\sum_{k=1}^{n} k^2 &= O(n^3), \quad \text{as} \quad n \to \infty; \\
\sum_{k=1}^{n} k^2 &= \frac{n^3}{3} + O(n^2), \quad \text{as} \quad n \to \infty; \\
\sum_{k=1}^{n} k^3 &= O(n^4), \quad \text{as} \quad n \to \infty; \\
\sum_{k=1}^{n} k^3 &= \frac{n^4}{4} + O(n^3), \quad \text{as} \quad n \to \infty. \quad \Box
\end{aligned}$$

Problem 11: Use the binomial formula

$$(a+b)^n = \sum_{j=0}^{n} \binom{n}{j} a^j b^{n-j}, \tag{10.33}$$

for all $a, b \in \mathbb{R}$, and for any positive integer $n \geq 1$, to establish the following equalities:

$$\sum_{j=0}^{n} \binom{n}{j} = 2^n; \tag{10.34}$$

$$\sum_{k=0}^{\lfloor \frac{n}{2} \rfloor} \binom{n}{2k} = 2^{n-1}; \tag{10.35}$$

$$\sum_{j=0}^{n} j \binom{n}{j} = n2^{n-1}. \tag{10.36}$$

Solution: By letting $a = x$ and $b = 1$ in the binomial formula (10.33), we obtain that

$$(x+1)^n = \sum_{j=0}^{n} \binom{n}{j} x^j, \quad \forall\, x \in \mathbb{R}. \tag{10.37}$$

If we let $x = 1$ in (10.37), we find that

$$2^n = \sum_{j=0}^{n} \binom{n}{j}, \tag{10.38}$$

which is the same as (10.34).

By differentiating (10.37) with respect to x, it follows that

$$n(x+1)^{n-1} = \sum_{j=1}^{n} \binom{n}{j} jx^{j-1}, \quad \forall\, x \in \mathbb{R}. \tag{10.39}$$

If we let $x = 1$ in (10.39), we find that

$$n2^{n-1} = \sum_{j=1}^{n} j \binom{n}{j},$$

which is the same as (10.36).

To prove that (10.35) holds true, let $x = -1$ in (10.37). Then,

$$0 = \sum_{j=0}^{n} (-1)^j \binom{n}{j}. \tag{10.40}$$

We add (10.38) to (10.40), and find that

$$2^n = \sum_{j=0}^{n} \left(1 + (-1)^j\right) \binom{n}{j}. \tag{10.41}$$

Note that

$$1 + (-1)^j = 2, \text{ if } j \text{ is even;}$$
$$1 + (-1)^j = 0, \text{ if } j \text{ is odd.}$$

Thus,

$$\sum_{j=0}^{n} \left(1 + (-1)^j\right) \binom{n}{j} = 2 \sum_{k=0}^{\lfloor \frac{n}{2} \rfloor} \binom{n}{2k}. \tag{10.42}$$

From (10.41) and (10.42), we conclude that

$$\sum_{k=0}^{\lfloor \frac{n}{2} \rfloor} \binom{n}{2k} = 2^{n-1},$$

which is the same as (10.35). □

Problem 12: Show that the series

$$\sum_{k=1}^{\infty} \frac{1}{k^2}$$

is convergent, while the series

$$\sum_{k=1}^{\infty} \frac{1}{k} \quad \text{and} \quad \sum_{k=2}^{\infty} \frac{1}{k\ln(k)}$$

are divergent, i.e., equal to ∞.

Solution: Since all the terms of the series $\sum_{k=1}^{\infty} \frac{1}{k^2}$ are positive, it is enough to show that the partial sums $\sum_{k=1}^{n} \frac{1}{k^2}$ are uniformly bounded, in order to conclude that the series is convergent. This can be seen as follows:

$$\begin{aligned} \sum_{k=1}^{n} \frac{1}{k^2} &= 1 + \sum_{k=2}^{n} \frac{1}{k^2} \leq 1 + \sum_{k=2}^{n} \frac{1}{k(k-1)} = 1 + \sum_{k=2}^{n} \left(\frac{1}{k-1} - \frac{1}{k} \right) \\ &= 1 + \left(1 - \frac{1}{n} \right) < 2, \quad \forall\ n \geq 2. \end{aligned}$$

To show that the series $\sum_{k=1}^{\infty} \frac{1}{k}$ is divergent, we will prove that

$$\ln(n) + \frac{1}{n} < \sum_{k=1}^{n} \frac{1}{k} < \ln(n) + 1, \quad \forall\ n \geq 1. \tag{10.43}$$

The integral of the function $f(x) = \frac{1}{x}$ over the interval $[1, n]$ can be approximated from above and below as follows: Note that

$$\int_1^n \frac{1}{x}\, dx = \sum_{k=1}^{n-1} \int_k^{k+1} \frac{1}{x}\, dx.$$

Since $f(x) = \frac{1}{x}$ is a decreasing function, it is easy to see that

$$\frac{1}{k+1} < \frac{1}{x} < \frac{1}{k}, \quad \forall\, x \in (k,\ k+1).$$

Then,

$$\begin{aligned} \int_1^n \frac{1}{x}\, dx &= \sum_{k=1}^{n-1} \int_k^{k+1} \frac{1}{x}\, dx > \sum_{k=1}^{n-1} \int_k^{k+1} \frac{1}{k+1}\, dx \\ &= \sum_{k=1}^{n-1} \frac{1}{k+1} = -1 + \sum_{k=1}^{n} \frac{1}{k} \end{aligned} \tag{10.44}$$

$$\begin{aligned} \int_1^n \frac{1}{x}\, dx &= \sum_{k=1}^{n-1} \int_k^{k+1} \frac{1}{x}\, dx < \sum_{k=1}^{n-1} \int_k^{k+1} \frac{1}{k}\, dx \\ &= \sum_{k=1}^{n-1} \frac{1}{k} = -\frac{1}{n} + \sum_{k=1}^{n} \frac{1}{k} \end{aligned} \tag{10.45}$$

The inequality (10.43) follows from (10.44) and (10.45), since

$$\int_1^n \frac{1}{x}\, dx \;=\; \ln(n).$$

In a similar fashion, by considering the integral of $\frac{1}{x\ln(x)}$ over the interval $[2,\ n]$, we can show that

$$\ln(\ln(n)) - \ln(\ln(2)) \;+\; \frac{1}{n\ln(n)} \;<\; \sum_{k=2}^{n} \frac{1}{k\ln(k)}, \qquad \forall\, n \geq 2; \tag{10.46}$$

$$\sum_{k=2}^{n} \frac{1}{k\ln(k)} \;<\; \ln(\ln(n)) - \ln(\ln(2)) \;+\; \frac{1}{2\ln(2)}, \qquad \forall\, n \geq 2, \tag{10.47}$$

and conclude that the series $\sum_{k=2}^{\infty} \frac{1}{k\ln(k)}$ is divergent.

For example, (10.46) can be proved as follows:

$$\int_2^n \frac{1}{x\ln(x)}\, dx \;=\; \sum_{k=2}^{n-1} \int_k^{k+1} \frac{1}{x\ln(x)} dx < \sum_{k=2}^{n-1} \int_k^{k+1} \frac{1}{k\ln(k)} dx = \sum_{k=2}^{n-1} \frac{1}{k\ln(k)},$$

which is equivalent to

$$\int_2^n \frac{1}{x\ln(x)}\, dx \;+\; \frac{1}{n\ln(n)} \;<\; \sum_{k=2}^{n} \frac{1}{k\ln(k)}. \tag{10.48}$$

Since

$$\int_2^n \frac{1}{x\ln(x)}\, dx \;=\; \ln(\ln(n)) - \ln(\ln(2)), \tag{10.49}$$

we conclude from (10.48) and (10.49) that

$$\ln(\ln(n)) - \ln(\ln(2)) \;+\; \frac{1}{n\ln(n)} \;<\; \sum_{k=2}^{n} \frac{1}{k\ln(k)},$$

which is the same as (10.46). □

Problem 13: Show that the sequence

$$x_n \;=\; \left(\sum_{k=1}^{n} \frac{1}{k}\right) \;-\; \ln(n)$$

is convergent to a limit between 0 and 1.

Solution: Recall from (10.43) that

$$\ln(n) + \frac{1}{n} < \sum_{k=1}^{n} \frac{1}{k} < \ln(n) + 1, \quad \forall\ n \geq 1,$$

which can be written as

$$\frac{1}{n} < x_n < 1, \quad \forall\ n \geq 1.$$

It is easy to see that

$$x_{n+1} - x_n = \frac{1}{n+1} - \ln(n+1) + \ln(n).$$

Therefore, $x_{n+1} < x_n$ if and only if

$$\frac{1}{n+1} < \ln(n+1) - \ln(n) = \ln\left(\frac{n+1}{n}\right).$$

This is equivalent to $1 < (n+1)\ln\left(\frac{n+1}{n}\right)$, and therefore to

$$e < \left(\frac{n+1}{n}\right)^{n+1} = \left(1 + \frac{1}{n}\right)^{n+1},$$

which holds for any $n \geq 1$, from the definition of e.

We showed that the sequence $(x_n)_{n=1:\infty}$ is decreasing and bounded from below by 0 and from above by 1, The sequence is therefore convergent to a limit between 0 and 1. □

Problem 14: Find the radius of convergence R of the power series

$$\sum_{k=2}^{\infty} \frac{x^k}{k\ln(k)},$$

and investigate what happens at the points x where $|x| = R$.

Solution: It is easy to see that

$$\sum_{k=2}^{\infty} \frac{x^k}{k\ \ln(k)} = \sum_{k=2}^{\infty} a_k x^k, \tag{10.50}$$

with

$$a_k = \frac{1}{k\ln(k)}, \quad \forall k \geq 2.$$

Note that

$$\lim_{k\to\infty} |a_k|^{1/k} = \lim_{k\to\infty} \left(\frac{1}{k\ln(k)}\right)^{1/k} = 1. \tag{10.51}$$

Recall that, if $\lim_{k\to\infty} |a_k|^{1/k}$ exists, the radius of convergence of the series $\sum_{k=2}^{\infty} a_k x^k$ is given by

$$R = \frac{1}{\lim_{k\to\infty} |a_k|^{1/k}}. \tag{10.52}$$

From (10.51) and (10.52) the radius of convergence of the series (10.50) is $R = 1$ We conclude that the series is convergent if $|x| < 1$, and not convergent if $|x| > 1$.

If $x = -1$, the series becomes $\sum_{k=1}^{\infty} \frac{(-1)^k}{k\ln(k)}$. Since the terms $\frac{(-1)^k}{k\ln(k)}$ have alternating signs and decrease in absolute value to 0, the series is convergent.

If $x = 1$, the series becomes $\sum_{k=1}^{\infty} \frac{1}{k\ln(k)}$, which was shown to be divergent in Problem 12 of this chapter. □

Problem 15: Let $a > 0$ be a positive number. Compute

$$\sqrt{a + \sqrt{a + \sqrt{a + \ldots}}}$$

Solution: If we know that the limit of $\sqrt{a + \sqrt{a + \sqrt{a + \ldots}}}$ exists, and denote that limit by l, then it follows that l must satisfy

$$l = \sqrt{a + \sqrt{a + \sqrt{a + \ldots}}} = \sqrt{a + l}. \tag{10.53}$$

which can be solved for l to obtain that

$$l = \frac{1 + \sqrt{1 + 4a}}{2}. \tag{10.54}$$

We now show that, for any $a > 0$, the limit of $\sqrt{a + \sqrt{a + \sqrt{a + \ldots}}}$ does exist, which is equivalent to proving that the sequence $(x_n)_{n\geq 0}$ is convergent, where $x_0 = \sqrt{a}$ and

$$x_{n+1} = \sqrt{a + x_n}, \quad \forall\, n \geq 0.$$

We prove that the sequence $(x_n)_{n\geq 0}$ is bounded from above and is increasing.

Let l be given by (10.54), i.e., let

$$l = \frac{1 + \sqrt{1 + 4a}}{2}.$$

(i) The sequence $(x_n)_{n\geq 0}$ is bounded from above by l.
Note that $x_0 = \sqrt{a} < l$. If we assume that $x_n < l$, then

$$x_{n+1} \; = \; \sqrt{a + x_n} \; < \; \sqrt{a + l} \; = \; l,$$

since l is the positive solution of (10.53), and therefore $l = \sqrt{a + l}$. Thus, by induction, we find that $x_n < l$ for all $n \geq 0$.

(ii) The sequence $(x_n)_{n\geq 0}$ is increasing.
It is easy to see that

$$x_n < x_{n+1} \iff x_n < \sqrt{a + x_n} \iff x_n^2 - x_n - a < 0,$$

since $x_n > 0$. Note that

$$\begin{aligned} x_n^2 - x_n - a &= \left(x_n - \frac{1 + \sqrt{1 + 4a}}{2}\right)\left(x_n - \frac{1 - \sqrt{1 + 4a}}{2}\right) \\ &= (x_n - l)\left(x_n + \frac{\sqrt{1 + 4a} - 1}{2}\right) \\ &< 0, \end{aligned}$$

since $x_n < l$ and $x_n > 0$ for all $n \geq 0$. We conclude that $x_n < x_{n+1}$ for all $n \geq 0$.

The sequence $(x_n)_{n\geq 0}$ is increasing and bounded from above, and therefore it is convergent. Let $l = \lim_{n\to\infty} x_n$. Then, as shown above, l satisfies the equation $l = \sqrt{a + l}$ and is given by (10.54), i.e.,

$$\lim_{n\to\infty} x_n \; = \; \frac{1 + \sqrt{1 + 4a}}{2}. \quad \square$$

Problem 16: Let $a > 0$ be a positive number. Compute

$$a + \cfrac{1}{a + \frac{1}{a + \ldots}}. \tag{10.55}$$

Solution: If we know that the continuous fraction (10.55) has a limit l, then l must satisfy

$$l = a + \frac{1}{l} \iff l^2 - al - 1 = 0. \tag{10.56}$$

Note that equation (10.56) has two solutions,

$$l_1 = \frac{a + \sqrt{a^2 + 4}}{2} \quad \text{and} \quad l_2 = \frac{a - \sqrt{a^2 + 4}}{2}.$$

Since l must be positive, and since $l_2 < 0$, we obtain that

$$l = \frac{a + \sqrt{a^2+4}}{2}. \tag{10.57}$$

To show that the continuous fraction (10.55) does have a limit, we must prove that the sequence $(x_n)_{n\geq 0}$ is convergent, where $x_0 = a$ and

$$x_{n+1} = a + \frac{1}{x_n}, \quad \forall\, n \geq 0.$$

The first few terms of the sequence $(x_n)_{n\geq 0}$ are

$$x_0 = a; \quad x_1 = \frac{a^2+1}{a}; \quad x_2 = \frac{a^3+2a}{a^2+1}; \quad x_3 = \frac{a^4+3a^2+1}{a^3+2a}.$$

We note that the terms of the sequence are alternatively larger and smaller than the value of l given by (10.57), i.e.,

$$x_0 < x_2 < l < x_3 < x_1.$$

Based on this observation, we conjecture that the subsequence $(x_{2n})_{n\geq 0}$ made of the even terms of the sequence $(x_n)_{n\geq 0}$ is increasing and has limit equal to l, and that the subsequence $(x_{2n+1})_{n\geq 0}$ made of the odd terms of $(x_n)_{n\geq 0}$ is decreasing and has limit equal to l.

To show this, let $(y_n)_{n\geq 0}$ be the sequence given by the recursion

$$y_{n+1} = a + \frac{1}{a + \frac{1}{y_n}} = \frac{(a^2+1)y_n + a}{ay_n + 1}, \quad \forall\, n \geq 0, \tag{10.58}$$

with $y_0 = a$. Note that $y_n = x_{2n}$ for all $n \geq 0$.

Assume that $y_n < l$, where l is given by (10.57). Recall from (10.56) that

$$l^2 - al - 1 = 0 \tag{10.59}$$

and note that

$$\begin{aligned} t^2 - at - 1 &= \left(t - \frac{a+\sqrt{a^2+4}}{2}\right)\left(t - \frac{a-\sqrt{a^2+4}}{2}\right) \\ &= (t-l)\left(t + \frac{\sqrt{a^2+4}-a}{2}\right). \end{aligned} \tag{10.60}$$

We will show that, for all $n \geq 0$, $y_{n+1} > y_n$ and $y_{n+1} < l$.

Note that, by definition (10.58), $y_n > 0$ for all $n \geq 0$. Then, from (10.58), it is easy to see that

$$y_{n+1} > y_n \iff (a^2+1)y_n + a > ay_n^2 + y_n \iff a(y_n^2 - ay_n - 1) < 0. \tag{10.61}$$

From (10.60), and using the assumption that $y_n < l$, it follows that

$$y_n^2 - ay_n - 1 = (y_n - l)\left(y_n + \frac{\sqrt{a^2+4}-a}{2}\right) < 0. \tag{10.62}$$

From (10.61) and (10.62), we conclude that, if $y_n < l$, then $y_{n+1} > y_n$, for any $n \geq 0$.

From (10.58), we also find that

$$y_{n+1} < l \iff (a^2+1)y_n + a < aly_n + l \iff y_n < \frac{l-a}{a^2 - al + 1} = l; \tag{10.63}$$

the last equality can be derived as follows:

$$\frac{l-a}{a^2-al+1} = l \iff l - a = a^2 l - al^2 + l \iff a(l^2 - al - 1) = 0;$$

cf. (10.59).

We conclude from (10.63) that, if $y_n < l$, then $y_{n+1} < l$ for all $n \geq 0$.

In other words, we showed by induction that the sequence $(y_n)_{n\geq 0}$ given by the recursion (10.58) with $y_0 = a$ is increasing and bounded from above by l. Therefore, the sequence $(y_n)_{n\geq 0}$ is convergent. Denote by $l_1 = \lim_{n\to\infty} y_n$ the limit of the sequence $(y_n)_{n\geq 0}$. From (10.58) and using (10.60), we obtain that

$$l_1 = \frac{(a^2+1)l_1 + a}{al_1 + 1} \iff a(l_1^2 - al_1 - 1) = 0$$
$$\iff a(l_1 - l)\left(l_1 + \frac{\sqrt{a^2+4}-a}{2}\right) = 0.$$

Since $l_1 > 0$, it follows that $l_1 = l$, i.e., that $\lim_{n\to\infty} y_n = l$.

Recall that $y_n = x_{2n}$ for all $n \geq 0$. We showed that the subsequence made of the even terms of $(x_n)_{n\geq 0}$ is increasing and converges to the limit l given by (10.57), i.e., that

$$\lim_{n\to\infty} x_{2n} = l. \tag{10.64}$$

Similarly, we define the sequence $(z_n)_{n\geq 0}$ by the recursion

$$z_{n+1} = \frac{(a^2+1)z_n + a}{az_n + 1}, \quad \forall\, n \geq 0,$$

with $z_0 = \frac{a^2+1}{a}$. It is easy to see that $z_n = x_{2n+1}$ for all $n \geq 0$.

As expected, the sequence $(z_n)_{n\geq 0}$ is decreasing and has limit equal to l. The proof follows by induction: assuming that $z_n > l$, we show that

$z_{n+1} < z_n$ and $z_{n+1} > l$. This proof is very similar to that given above for the sequence $(y_n)_{n\geq 0}$ and is left to the reader. We conclude that

$$\lim_{n\to\infty} x_{2n+1} = l. \quad (10.65)$$

From (10.64) and (10.65), we find that

$$\lim_{n\to\infty} x_n = l = \frac{a+\sqrt{a^2+4}}{2}. \quad \square$$

Problem 17: (i) Find $x > 0$ such that

$$x^{x^{x^{\cdot^{\cdot^{\cdot}}}}} = 2. \quad (10.66)$$

(ii) Find the largest possible value of $x > 0$ with such that there exists a number $b > 0$ with

$$x^{x^{x^{\cdot^{\cdot^{\cdot}}}}} = b. \quad (10.67)$$

Also, what is the largest possible value of b?

Solution: (i) If there exists x such that (10.66) holds true, then

$$x^2 = x^{x^{x^{x^{\cdot^{\cdot^{\cdot}}}}}} = 2,$$

and therefore $x = \sqrt{2}$. We are left with proving that

$$\sqrt{2}^{\sqrt{2}^{\sqrt{2}^{\cdot^{\cdot^{\cdot}}}}} = 2.$$

Consider the sequence $(x_n)_{n\geq 0}$ with $x_0 = \sqrt{2}$ and satisfying the following recursion:

$$x_{n+1} = \left(\sqrt{2}\right)^{x_n} = 2^{x_n/2}, \quad \forall\, n \geq 0.$$

It is easy to see by induction that the sequence is increasing and bounded from above by 2, since

$$x_{n+1} > x_n \iff 2^{x_n/2} > 2^{x_{n-1}/2} \iff x_n > x_{n-1};$$

$$x_{n+1} < 2 \iff 2^{x_n/2} < 2 \iff x_n/2 < 1 \iff x_n < 2.$$

We conclude that the sequence $(x_n)_{n\geq 0}$ is convergent. If $l = \lim_{n\to\infty} x_n$, then

$$l = 2^{l/2},$$

which is equivalent to

$$l^{1/l} = 2^{1/2}. \quad (10.68)$$

Let $f : (0, \infty) \to (0, \infty)$ be given by

$$f(t) = t^{1/t} = \exp\left(\frac{\ln(t)}{t}\right).$$

Then

$$f'(t) = \frac{1 - \ln(t)}{t^2} \exp\left(\frac{\ln(t)}{t}\right).$$

Note that the function $f(t)$ is increasing for $t < e$ and decreasing for $t > e$. Therefore, there will be two values of t such that (10.68) is satisfied, i.e., such that $t^{1/t} = 2^{1/2}$, one value being equal to 2, and the other one greater than e. Since $x_n < 2$ for all $n \geq 0$ and $l = \lim_{n\to\infty} x_n$, we conclude that $l = 2$, and therefore that $x = \sqrt{2}$ is the solution to (10.66).

(ii) If there exists a number $b > 0$ such that

$$x^{x^{x^{\cdot^{\cdot^{\cdot}}}}} = b$$

for a given $x > 0$, then $x^b = b$ and therefore $x = b^{1/b}$. Recall from part (i) that the function $f(t) = t^{1/t}$ has an absolute maximum at $t = e$. Thus,

$$x = b^{1/b} \leq \sup_{t>0} t^{1/t} = e^{1/e} \approx 1.4447.$$

We conclude that the largest value of x such that the limit (10.67) exists is $x = e^{1/e}$. The largest value of b such that the limit (10.67) exists is $b = e$. □

Bibliography

[1] Dan Stefanica. *A Mathematical Primer with Numerical Methods for Financial Engineering.* FE Press, New York, 2nd edition, 2011.

Made in the USA
Columbia, SC
18 September 2023